A
Scientist
at the
Seashore

James Trefil

Illustrations by
Gloria Walters

DOVER PUBLICATIONS, INC.
Mineola, New York

Bibliographical Note

This Dover edition, first published in 2005, is an unabridged, slightly corrected republication of the work as published by Collier Books, Macmillan
Publishing Company, New York, in 1987. The work was originally published
in 1984 by Charles Scribner's Sons, New York.

Library of Congress Cataloging-in-Publication Data

Trefil, James S., 1938–
    A scientist at the seashore / James Trefil ; illustrations by Gloria Walters.
       p. cm.
    Originally published: New York : Scribner, c1984.
    Includes index.
    ISBN 0-486-44564-X (pbk.)
       1. Physics. 2. Beaches. I. Title.

QC28.T67 2005
530—dc22

                                                                    2005048369

Manufactured in the United States of America
Dover Publications, Inc., 31 East 2nd Street, Mineola, N.Y. 11501

CONTENTS

INTRODUCTION

The science of physics, in addition to giving us the ability to describe and predict the behavior of all objects from galaxies to electrons, also provides us with a unique view of the world. The key factor in the physicist's outlook is the concept of natural law. When the period of turmoil that characterizes the birth of a new science is over, the physicist expects to find a set of basic regularities that hold true in the area being studied. For example, in 1687 Isaac Newton summarized the principles of mechanics (the area of science concerned with the motion of material bodies) in his three laws of motion. Everything that had to do with motion, from the orbits of planets to the fall of an apple, was contained in these three simple laws. During the nineteenth century, two more areas of nature were explained in corresponding fashion: heat in (three laws of thermodynamics) and electricity and magnetism (in four laws called the Maxwell equations). In this century gravitation (one general principle of relativity) and quantum mechanics (four or five basic postulates, depending on which side you take in some ongoing debates) have been added to the list.

The lesson to learn from these developments is this: If everything that happens in the world can be organized on the basis of only a few natural laws, then many apparently different phenomena must be connected with each other. In fact, the world as seen by the physicist is like a large interconnected web. Starting with what is immediately available to our senses, this web must become more and more constricted as we follow it, until finally every strand ends at one of the great physical principles mentioned above.

From this view of nature an important conclusion can be drawn. If every phenomenon we encounter ultimately leads us to one of the general laws, then it makes no difference where we start. *Anything* can serve as the starting point for an inquiry into the workings of the physical world. You don't have to start in a laboratory because, in a very real sense, the entire world is a physics laboratory. So important is this truth that the present book is entirely dedicated to the exploration of one site, a beach, for clues to the ultimate workings of the universe.

We shall see that the ebb and flow of the tide is connected with the fact that no one on the surface of the earth can ever see the back side of the moon, and related as well to the search for new planets in the solar system—a search which may not be over even today. Waves and surf are related to new technologies for communication and to research into the past climates of the earth. Something as ephemeral and inconsequential as a bubble in the foam leads us to consider the forces that hold the nucleus of the atom together. Even an ordinary sailboat has its tale to tell and its lessons to teach. At every turn, we find new corroboration for the beautiful and elegant concept of natural law.

Before we step down to the beach, let me say a few words more about the physicist's viewpoint you will find in this book. You will perhaps be struck first by the fact that all the topics I choose to discuss deal with inanimate objects—waves, sand, stones, and so on. This isn't because I don't see and enjoy the wildlife along the beach. I do. But *as a physicist*, I have nothing unusual to say about it. Because I am not trained in zoology or botany, I simply can't see the web to which sand crabs and gulls are connected, nor have I anything more than a layman's knowledge of the principles which made them the way they are. I have, therefore, chosen to confine myself to subjects where my own special training is of use. This is but to follow Emerson's old dictum, "Tell us what you know."

The second point has to do with an attitude I sometimes encounter in people with whom I've discussed scientific ideas. They argue that understanding how something works somehow diminishes our appreciation of its beauty. I particularly recall one author I read while a graduate student. He quoted some lines from Keats about the beauty of sunlight on water and contrasted them with a passage dealing with the formation of ripples from a textbook on hydrodynamics. I have to admit that I was irritated by this example of a non sequitur, particularly since I read it while sitting on the shore of Lake Lagunita on the Stanford campus, where I regularly spent my lunch breaks reading and enjoying the sunlight on water. But personal feelings aside, I have never been able to see how understanding something makes it less beautiful. An understanding of the behavior of light doesn't detract from the beauty of a Botticelli painting, nor does understanding the role of stress and strain in the structure of a cathedral make the experience of walking into one any less compelling. To my mind, deeper understanding makes experience richer. A beach—perhaps one of the most pleasant of natural environments—is no exception. A scientist walking on the sand sees the same things there that anyone else does. The fact that he knows more about some portion of what he sees does not diminish his powers of appreciation nor the quality of his enjoyment.

With this assurance, I invite you to walk along your favorite beach with me, deepening your appreciation of the experience by seeing some of it through my eyes.

JAMES TREFIL
Charlottesville, Virginia

process started to occur. When a meteor hits anyting, some material sticks and some is scattered back into space by the impact. Of this scattered material, some is moving fast enough to go back into orbit and some is eventually pulled back to the planet. The lower the density of the material, the more likely it is to escape. Thus, in the early stages of its formation, the earth collected heavy material, leaving lighter stuff (such as silicon and water) still in orbit around the sun.

As the earth began to approach its present size, however, the ever-growing gravitational attraction made it harder and harder for debris to escape. By this time, most of the material in the vicinity of the earth's orbit consisted of debris of previous collections and was therefore composed largely of lighter elements. The earth swept through this material, picking up mass the way the windshield of a moving car picks up insects on a summer afternoon. The silicon that makes up the sand on the beach, as well as the water that makes up the ocean, were added to the earth during this last phase of accretion. Each bit of sand or water that you see, therefore, has probably been in space more than once during its history.

The formation of the earth from planetesimals and the sun from the bulk of the original gas cloud took place simultaneously and required a few hundred million years to be completed. When the earth was forming, the fusion reaction in the sun ignited. It was not a smooth process, and can be likened to starting a car engine on a cold morning. The sun stuttered, balked, and backfired for a while—a situation astronomers describe as the *T Tauri* stage in stellar evolution. Each flaring up of the sun sent strong streams of particles rushing out. These particles swept the remaining gases out of the inner solar system, leaving it in much its present state. Had the earth been formed with a readymade atmosphere present, it would have been blown off at this time, leaving the early earth as a rocky ball with neither air nor water on its surface.

But if all of the water in the atmosphere had been blown out of the solar system by the sun's flare-up, where could the oceans have come from? The answer, paradoxical as it may seem, is that they came from inside the earth. The outer layer of the planet was rich in light materials, including water which had been swept up. The new world was a turbulent place, with volcanoes and earthquakes and other types of what we now call tectonic activity. Each time a volcano erupted or a geyser went off, material from the interior (primarily water vapor but including all sorts of gases) entered the new atmosphere. The process of releasing volatiles in this way is called degassing. Over a relatively short time—100 million years or so—enough volatile material had been released in this way to form the oceans and to give the earth an atmosphere. It wasn't an atmosphere like our own; there was almost no free oxygen, for one thing. But it was a collection of gases which were held at the planetary surface by gravity and which therefore deserves the title of an atmosphere.

Early in this history, perhaps almost from the beginning, the temperature of the earth fell below 212° Fahrenheit and the water condensed into oceans as we know them today. This is not to say that our present oceans formed 4.5 billion years ago—far from it. It simply implies that the large amounts of water we associate with an ocean were on the scene within a few hundred million years of the earth's creation. As we shall see in the next chapter, the present ocean structure is fairly new, the Atlantic Ocean being only 165 million years old.

The weight of geological evidence points to another amazing fact about the early oceans. The amount of water contained within the oceans has not changed appreciably since they were formed. The mass of water in the oceans now (about 10^{24} gm) is roughly the same as the mass of water that was contained in the crust of the earth when the degassing started. Furthermore, we can estimate the rate at which water is being lost today. The water molecule is too heavy to escape the earth's gravitational pull easily, but water molecules in the air are occasionally dissociated (broken up into the constituent hydrogen and oxygen). The hydrogen freed in this way is light enough to move off into space, which is why there is so little free hydrogen in the earth's atmosphere. The net effect of hydrogen loss, then, is to decrease the amount of water vapor in the atmosphere—water vapor which is replaced by evaporation from the oceans.

About 5×10^{11} gm of water are lost this way each year. This corresponds to a cube 100 yards to a side—about the volume of a small lake. All the water lost to space since the beginning of the earth amounts to about 2×10^{21} gm—less than 0.2 percent of the water in the oceans. In fact, all of the water lost to space since the beginning corresponds to a square of ocean about eight hundred miles on a side. This means that most of the water you see when you stand on the beach is, in fact, the very same stuff that was degassed from the interior when the earth was only a few hundred million years old.

The small amount of water lost to space is replaced by the same geological processes that formed the present ocean basins. In the next chapter I shall discuss the process by which new ocean floor is created by material moving up to the surface, pushing the old floor aside. This upwelling material contains some water, which, when added to the oceans, balances the loss to space. Geologists call this material added to the oceans "juvenile water." (When I see a scientific term as marvelous as this, I weep for my own field, elementary particle physics, whose contributions include terms like "quark" and misnomers like "strangeness" and "color." For more on this, see my book *From Atoms to Quarks.*)*

In any case, the balance between water lost to space and juvenile water added from the interior guarantees that the total amount of water on the earth's surface must be roughly constant over long periods of time. It does not, however, mean that all that water must be in liquid form. During the ice ages, an appreciable portion of the water now in the oceans was locked up in glaciers. If there is a fixed amount of water on the surface, then it follows that glaciers can grow only at the expense of the oceans. Thus, when ice covered most of North America and Europe eighteen thousand years ago, the ocean levels were lower than they are now. Large portions of the North Sea and the Baltic Sea were dry land, and Britain was attached to mainland Europe. The east coast of the United States was a hundred miles farther east than it is now, and what is currently part of the ocean floor was populated by land animals (including man).

In a similar vein, the present coastlines of the world are where they are because some of the water supply is tied up in the polar ice caps, mainly on Antarctica. Should the ice caps melt so that all of the water flowed into the ocean basins, the sea level would increase by several hundred feet and flood most of the low-lying coastal areas of the world. But in any case, the fact that your favorite beach is where it is and not a few hundred miles out to sea (or inland) is the result of the way the earth's total water supply is divided between the liquid in the ocean and the solid in glacier and ice cap.

* New York: Anchor Books, 1994 (second edition).

Having found the origins of the ocean in the processes that formed the earth, we can turn to the second (and more difficult) problem we posed earlier. Given that there was enough water to form the oceans early in the earth's history, how is it that the water is still here now, some four billion years later? Presumably the scheme outlined above, in which a planet degasses after its formation, could have happened on any of our neighbors in the solar system, including the moon. Why, then, has the earth alone retained its oceans?

The case of the moon is easy to understand. Because of its light mass, atoms in the moon's early atmosphere had an easy time escaping into space. Long ago they slipped off, leaving the surface exposed to vacuum as it is today. The case of the planets is a little more complicated, so that to understand the present state of the earth we must turn to the remarkable concept of the "continuously habitable zone" (CHZ) introduced by astrophysicist Michael Hart in 1978.

But first we have to realize that during the lifetime of the earth a number of major changes have taken place in the environment, any one of which would be potentially disastrous for the continued existence of the ocean. To begin with, the luminosity of the sun has increased by about twenty-five percent between the time the earth was formed and the present. Had this happened without any compensating changes in the atmosphere, the oceans would have boiled away long ago. Similarly, about two billion years ago oxygen produced by photosynthesis in algae in the earth's oceans started to accumulate in the atmosphere. The oxygen destroyed several gases that were present in the atmosphere by reacting chemically with them. The most important of these were ammonia and methane. Both of these compounds contributed to the greenhouse effect—they were part of the atmospheric "blanket" that kept the earth warm. Removing them was like tearing holes in that blanket; heat started streaming out and the earth cooled off rapidly—almost catastrophically, in fact. Had the cooling been just a little more violent and the energy coming in from the sun just a little less strong, the earth would have frozen solid as soon as a significant amount of oxygen had accumulated in the atmosphere.

What Hart found was that there is a very narrow band around the sun—a band which he called the CHZ—in which it is possible for a planet to tread the narrow line between boiling and freezing. Venus, for example, probably never had an ocean because it is so close to the sun. The atmosphere formed by degassing, acting through the greenhouse effect, kept the planet so warm that the temperature never dropped to the point where the water could condense. New gases added to the atmosphere simply stayed there, increasing the blanket effect and raising the surface temperature still higher. Mars, on the other hand, presents the opposite picture. Because of its small mass and great distance from the sun, whatever liquid water it once may have had has long since either escaped or frozen, leaving the arid ball we see today.

The existence of the CHZ, then, tells us where a planet must be placed relative to its sun in order for it to keep liquid water on its surface for the billions of years necessary for the evolution of life. In our solar system, the CHZ extends from a radius one percent larger than that of the earth's orbit to a radius five percent smaller. In other words, had the random process leading to the formation of the earth begun one percent farther away or five percent closer to the sun than it actually did, no higher life forms would ever have developed on the planet at all, and the earth today would either resemble Venus—the victim of a runaway greenhouse effect—or have been frozen solid for the past two billion years. This is what I meant earlier when I referred to the oceans of the earth as unique.

The same sort of reasoning that led to these rather stringent limits on where a planet can be located and still possess oceans can also be applied to the properties of the earth and the sun themselves. For example, were the mass of the sun less than eighty-three percent of what it is, any planet would freeze when oxygen accumulated in its atmosphere; there simply wouldn't be enough heat coming in to counteract the formation of glaciers. On the other hand, if the mass of the sun had been twenty percent larger than it is, it would have burned up all of its nuclear fuel before a billion years had elapsed. Consequently, only stars close in mass to the sun—what astronomers call G stars— have a CHZ and sufficient lifetime so that planets could have oceans like our own. Large stars have a large CHZ, but live for relatively short times. Small stars have no CHZ, but have long lifetimes. For liquid water to remain on a planet's surface for several billion years, then, requires a delicate balance between longevity and energy output in a planet's star.

Similarly, if the earth had been ten percent more massive than it is, the material emitted in the degassing process would have contained enough carbon dioxide to produce a runaway greenhouse effect, regardless of the type of star involved. Had the earth had ninety-four percent of its present mass, not enough of an ozone layer would have been formed to shield the surface of the earth from the sun's ultraviolet radiation, and life—if it developed at all—could never have emerged from the sea.

So the mere fact that we can stand on a beach and look out at the sea today, some 4.5 billion years after the formation of the earth, tells us that the earth is a unique place. It is unique not only in relation to our own solar system but in relation to the entire Milky Way galaxy. It is possible for you to visit the beach only because you live on a planet which is neither too large nor too small, circling at just the right distance from its star, which is also neither too large nor too small. And all of these conditions had to be met before the planet could produce a being capable of thinking about everything implied by standing on a beach.

Hart's discovery of the CHZ is typical of a major (and unremarked) trend in recent scientific research. When the question of extraterrestrial intelligence (ETI) first became scientifically respectable in the late 1950s, relatively little was known about the evolution of the solar system or the early earth. It was assumed that since the same laws of nature operate everywhere in the universe, the earth and its environs must be rather typical of what we would find if we could visit other parts of the galaxy. This is one reason why the presumption of the existence of water on Mars and Venus was so widespread. This idea, called the assumption of mediocrity, led to the well-publicized conclusions that there must be many forms of intelligent life in the universe, with the human race being a very junior member of the galactic club.

But the last twenty years have seen an enormous expansion of our understanding about the earth and the life upon it, and how both came into existence. The concept of the CHZ is one example of this new kind of understanding. It is typical of much of our new knowledge in showing that although the *laws* which operate on the earth are the same as they are everywhere in the universe, the *conditions* under which those laws operate are not. The earth, because of its size, its position, and the kind of star around which it circles meets a special and exacting set of criteria. It is unusual because it is one of the very few places in the galaxy where oceans can exist.

That may not seem an extraordinary discovery—until you remember that life on earth developed in the oceans. It is pretty clear that if intelligent life is to develop elsewhere from a carbon chemistry based on something like DNA,* it, like us, will have to develop in water. Hence, the statement that the earth is one of the few places where oceans can exist is equivalent to the statement that the earth is one of the few places where intelligent life can develop. This is enough to make the earth very special indeed.

So the next time you wander along your favorite beach, you might pause for a moment to reflect on the enormous improbability of the ocean, the beach, or you being there at all. You might also spare a moment to think about how this highly unlikely and very precious resource can be preserved for our children.

---

* The reasons why this should be so, and the arguments as to why all life must be carbon-based, are given in *Are We Alone*, by R. T. Rood and myself (New York: Charles Scribner's Sons, 1981).

2

THE SALT
SALT SEA

*According to an Old Norse folktale the sea is salt because some-
where at the bottom a magic salt mill is grinding away. The
tale is perfectly true. Only the details need to be worked out.*

—OCEANOGRAPHER FERREN MACINTYRE

The sea is salty, no doubt about it. All you have to do to to be sure of
that is put your finger in the water and taste. About three percent of
seawater is made up of minerals of one kind or another, with the largest
fraction—about nine-tenths of the total—being plain old everyday table
salt, a combination of the elements sodium and chlorine. Where these
two components of the salt in the sea came from and how they came
to dominate the chemistry of the ocean are, for all their apparent
simplicity, exceedingly deep questions.

The first scientific attempt to explain the origin of the salt in the
ocean was made by the Anglo-Irish scientist Robert Boyle in the 1670s.
Drawing on his studies of the chemistry of the atmosphere and the

11

phenomena of color, he made measurements which showed that rivers carried minute amounts of salt with them as they flowed into the sea. This led to a view of the origin of sea salt which served as the standard explanation for centuries—indeed, it is the one I was taught in grade school. It is, that minerals are continuously leached from rocks and soil by rainwater, and eventually join the runoff which forms local streams, move on to the rivers, and thence into the sea. Each drop of water that enters the sea brings with it some minerals from the land, and these minerals stay behind when the drop of water eventually evaporates and becomes rain again. In this picture, the river systems of earth form a kind of continuous conveyor belt, washing materials into the sea and leaving them behind as the water starts its evaporation-precipitation cycle again. The picture explains why the ocean is salty while rivers are fresh, since the amount of dissolved material carried in any given river is small. Only over a long period of time would you expect the content of dissolved minerals in the ocean to build up.

Unfortunately, this simple explanation of the ocean's chemical contents just won't do. For one thing, it is relatively simple to calculate how long it would take for the sea to reach its present level of salinity once you estimate the amount of salt brought into the oceans by rivers. The answer turns out to be surprisingly short—somewhat less than 100 million years. Since this is much less than the actual age of the earth, there is a clear contradiction.

Another problem has to do with an assumption implicit in this explanation. If the oceans are simply the passive receptacle of materials leached from the land, then they must be getting saltier as time goes by. In point of fact, studies of ancient sediments show that the concentration of salt in the sea was approximately the same 200 million years ago as it is today, despite all the water that has flowed into it. The data we have shows that the oceans have retained a three percent salinity since their origin.

Finally, geologists have pointed out that if the salt in the sea were simply supplied by river runoff, we would soon run out of land. For example, if calcium were to be transported to the sea at the presently measured rate and not replaced, in 100 million years all the fossil limestone on dry land (the main source of calcium in the rivers) would be exhausted. Again, we find that regarding the oceans as a giant waste receptable simply doesn't square with the facts.

The truth is, of course, that the former simple view of the role of the sea fails to take account of what happens to materials *after* they enter the oceans. Instead of picturing the sea as a huge kettle which passively acepts whatever is added to it, we should picture it as a large chemical laboratory, one in which the rivers do indeed provide the raw materials, but in which they are transformed by chemical reaction into what we actually find. Only by studying the active nature of marine chemistry can we hope to understand why the sea is salty and why the salinity seems to stay constant.

Perhaps the best way to make this point is to compare the amounts of some typical materials in the sea with estimates of the amounts being brought by rivers. This is done in the table below:

Concentrations of Elements in Water
(in gm/kg or parts per thousand)

material and chemical symbol	seawater content	net river content
sodium (Na^+)	10.7	.002
chlorine (Cl^-)	19.7	0
bicarbonate (HCO^-_3)	1.4	.06
magnesium (Mg^+)	1.2	.003
potassium (K^+)	.4	.002
calcium (Ca^{++})	.4	.015

All chemicals in the table actually exist as ions (rather than being electrically neutral) in solution. The plus and minus signs signify the electrical charge of the atoms in solution. Adapted from Peter Weyl, *Oceanography*. New York: John Wiley & Sons, 1970.

Obtaining the contents of seawater is fairly straightforward, but the river contents require two pieces of information. First, we have to have some idea of the concentrations of various elements being carried into the oceans. This is not a simple matter, since some rivers, like the Colorado, are very salty when they reach the ocean, while others, such as the Columbia, are not. The total input of material has to be averaged over the world's rivers.

More important, some of the elements the rivers carry with them actually came from the oceans in the first place. For example, small droplets of seawater in spray can remain suspended in the atmosphere long enough for them to dry out, which means that tiny windborn crystals of salt are suspended in the atmosphere. The presence of this suspended salt is what enables you to smell the sea from some distance inland. The salt is washed out of the air by the same rain which eventually flows into the rivers emptying into the sea. So to find what materials from the land are being carried out to sea, we have to subtract from the actual amount of each element transported some estimate of the contribution of ocean spray. This is why I have labeled the second column in the table *net* river content. In point of fact, there is a lot of chlorine in river water, but if we compare the measured amount to what would be contributed by the evaporation process, we find that the two figures are about equal. We conclude that almost all the chlorine is supplied to rivers by sea spray.

A glance at the table shows two important facts: first, the concentration of elements in river water is much lower than it is in seawater, as you might expect. Less obvious, however, is the fact that the *mix* of elements in seawater is radically different from the mix of elements carried by rivers. Seawater is primarily a solution of ordinary table salt (sodium chloride), but river water seems to be primarily a solution of calcium bicarbonate, with a relatively slight admixture of sodium. If we needed further evidence that the sea is not simply a passive receptacle for detritus from the weathering of rocks and soil, it is contained in the table.

How does the sea convert the elements from the river into salt? For each element in the table (and for others whose lower concentrations make them less important) there is a separate story. Potassium, for example, combines with various clays and rocks on the ocean bottoms to form claylike minerals which, over geological time, are converted to granite by heat and pressure. Calcium, on the other hand, is taken up by organisms in the ocean to form skeletons and shells. When the organisms die, these shells fall to the bottom and eventually are turned into the limestone and dolomite rocks that form such structures as the white cliffs of Dover and the Maritime Alps in southern France. The important chemical compound in this process is calcium carbonate $(CaCO_3)$.

The slow descent of calcium carbonate into the deep ocean actually produces one of the most picturesque effects in oceanography—the lysocline, or "snow line." The amount of calcium carbonate that can be dissolved in seawater depends strongly on pressure. At high "altitudes" above the ocean floor, the pressure is low and the calcium carbonate remains as a white solid. The shells and other detritus descend to form sediment, coating the ocean bottom with a layer of white and looking for all the world like snow. When the ocean depth is greater than about 14,000 feet, however, the pressure is high enough to dissolve the skeleton back into the water—though it may take many years to do so. Accordingly, an undersea mountain will be bare on its lower regions, but as soon as the side rises to about 14,000 feet below sea level, the skeletons remain undissolved and the white coating appears. The mountain will have a white cap, just like the Himalayas. By looking for the presence or absence of calcium carbonate in old sediments, then, oceanographers can learn something of the history of the ocean.

Another element, magnesium, undergoes much the same reaction as calcium, being taken into the very same shells as a minor component. Sodium will be taken up into rock formation by processes similar to those that affect potassium, although it takes longer for this to happen.

Thus, for most of the materials that are brought into the sea in significant quantities we have been able to ascertain some series of steps—either biological or geological—by which they are removed from the water. The average time that a given atom can remain in solution in the ocean is called the residence time for that element. The faster the chemical reactions take place, the shorter the residence time. Sodium, for example, has a residence time of some 68 million years, while for calcium the number is only about 1 million years. So if we put equal amounts of sodium and calcium into the ocean, in a million years most of the calcium would be on the ocean floor in the form of sediment, while ninety-eight percent of the sodium would still be in solution. It follows that the high concentration of sodium in the ocean is due to the fact that the chemical reactions that remove this particular element from the water proceed very slowly.

The other half of the salt molecule, chlorine, has a completely different history. We have already seen that all of the chlorine in river water can be said to originate in sea spray (some minor exceptions to this general rule will be mentioned later). This means that the total amount of chlorine in the sea, like the total volume of the water itself, must have been roughly all the same throughout the history of the earth. And, like the water in the sea, there is only one place the chlorine could have originated: in the interior of the earth. The chlorine in the ocean, like the earth's atmosphere, is a product of the degassing of the earth. Once in the sea, the residence time of the chlorine is very long. For all practical purposes, it is infinite. It leaves the sea for short periods of time in the form of airborne spray, giving rise to such phenomena as the salty rain that falls over or near the ocean. The predominant chlorine cycle is sketched in figure 2-1.

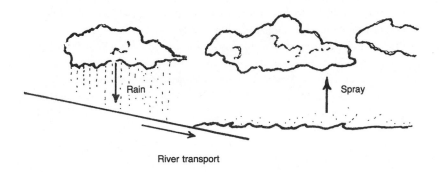

Figure 2-1.

The saltiness of the sea, then, is not so much a question of what elements are emptied in as it is a function of how long those elements stay there before they are removed by chemical reactions. Sodium and chlorine have long residence times in the water. One way to think about residence times is to realize that when you taste seawater, the sodium atom in the salt was probably deposited in the ocean when dinosaurs were still wandering the earth, while the chlorine was there from the beginning, over four billion years ago.

When I argued that the oceans should not be regarded as passive receptacles of land-originated waste material, I pointed out that it would take only 100 million years or so for weathering to transport all the material now above sea level to the ocean floor. The fine points of marine chemistry we've been discussing do not change this fact. Unless we can understand how the transportation of material to the ocean does not obliterate the land, we cannot claim to understand the workings of the sea.

There is a hint as to how we might proceed in the chlorine cycle.

We saw that the total amount of chlorine remained constant, but that separate atoms went through a cycle in which they were removed from the ocean (by spray) and subsequently returned to it (by rain). Is it possible that a similar cycle exists for sodium and other materials?

As we saw, the end point for most of the reactions that take place in the ocean is the incorporation of river-borne material into sea floor sediment. The only way to avoid the problem of the destruction of the continents would be to find some method of recycling this material—to create new land as the old weathers away.

The theory of plate tectonics provides at least a partial understanding of the process by which sodium and other elements can be returned to the land once they have been deposited as sediment. The theory is based on the idea that the surface of the earth is broken up into a dozen or so disjointed regions called plates. At some of the boundaries between plates, molten magma rock rises from the interior of the earth. As the magma emerges and cools, it pushes the plates (on which the continents ride) ahead of it. When two plates collide, the ocean floor will generally be pushed under the relatively more stable plate associated with a continent, resulting in an area of very high volcanic activity or in a deep ocean trench. The west coast of South America is such a region. This process is called *subduction*. At the subduction zones, the old ocean floor, with all of its sediments, is returned to the earth and remelted.

Geologists estimate that enough material is subducted to replace the entire crest of earth about every 100 million years. The actual process by which materials are subsequently returned to the land is poorly understood at present, and constitutes a major area of research. But if we accept plate tectonics, we have a satisfactory picture of the cycle followed by sodium. The floor of the ocean is actually a huge conveyor belt. It creates new floor at the rate of an inch a year or so, pushing old floor along as it does so. The older floor collects the sediments resulting from marine chemistry and carries them to subduction zones, where the conveyor belt and everything on it is melted.

The return trip of the belt, deep in the earth, is the part of the cycle which at the moment we don't understand. The tentative picture does solve our problem, though, because it shows that there is a way to recycle sodium and other materials. Instead of being cycled through the air like chlorine, they are cycled through the earth's mantle. A sketch of the sodium cycle is shown in figure 2-2. It should also be mentioned in passing that a small proportion of chlorine in the ocean follows this pattern, being taken up into minerals on the ocean floor and then returned, either in the upwelling magma or in the eruptions of volcanoes along the subduction zones.

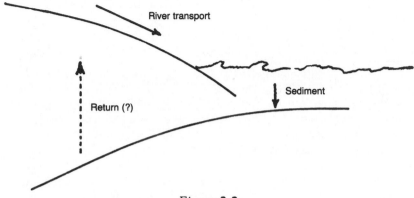

Figure 2-2.

In a sense, the ocean floor conveyor belt plays the role of the magic salt mill in the old Norse legend, grinding way forever to keep the sea salty. The only difference is that the tectonic salt mill actually removes sodium from the sea, rather than adding it.

One way to understand the importance of the creation of new sea floor is to consider the history of a particular ocean, the Atlantic. That ocean did not as yet exist 165 million years ago because the continents of North and South America, which now bound it on the west, and the continents of Europe and Africa, which now bound it on the east, were joined together. Although the amount of water in the world ocean was about the same, none of it was where the Atlantic Ocean is now. But molten magma, rising from the interior of the earth, caused the single land mass to separate, and the continued upwelling of new rock pushed the continents ever farther apart. The process is still going on today, as our dynamic picture of the ocean floor shows. In fact, one way of picturing the motion of the continents is to imagine them as sitting on plates of rock about sixty miles thick, with the plates floating around on the liquid rock of the lower mantle. Sea floor spreading— the same mechanism that recycles the salt in the sea—provides the motive force which keeps the plates and the continents on them in constant motion.

When the theory of continental drift was first put forward as a serious scientific proposal by the German meteorologist and arctic explorer Alfred Wegener in 1912, the mechanism for moving continents about was unknown. Wegener proposed the theory on geological and biological grounds, the most compelling of which is also the easiest to understand. A glance at a map of the Atlantic shows that the west coasts of Europe and Africa and the east coast of the Americas fit together like pieces of a jigsaw puzzle. The fit is even better if we look at the edges of the continental shelf under the sea instead of the present visible coastline. All this was suggestive, but it was not hard evidence for continental drift. It took almost half a century before a descendant of Wegener's original proposal became the main working hypothesis of earth scientists.

In the history of the Atlantic, the mechanism of plate tectonics worked itself out in three phases. At first, the split between the plates was shallow—perhaps only a mile or so in depth. The chasm created as the continents separated was largely filled with sediments carried into the new body of water by rivers. This stage is shown in figure 2-3. With the passage of time, the character of the ocean began to change. As the plates separated, magma flowed into the gap from underneath and solidified, forming the rocky floor of the new sea. Materials carried by rivers spilled down the steep inclines at the edges of the rift, forming the continental shelves. In addition, the calcium carbonate "snow" we mentioned earlier began to fall, so that the bottom was covered with a layer of mixed calcium and clay, the latter being brought in by the rivers. The Atlantic at age 40 million years (i.e., as it was 125 million years ago) is shown in figure 2-4.

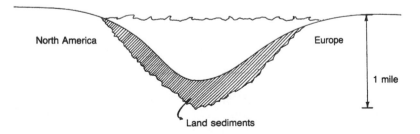

Figure 2-3.

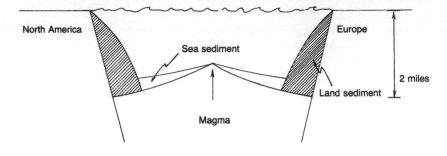

Figure 2-4.

As the continents continued to separate at the pace of an inch a year, the ocean basin began to take on its characteristic modern shape. The high point at the center of the ocean basin—the point which shows where the upwelling magma is forming new ocean floor—remains at roughly the same depth, about 7,500 feet. This mid-ocean ridge extends from Iceland to a point about a thousand miles off the coast of Antarctica and comprises a mountain chain greater than the Himalayas. The continental shelves extend a hundred miles or more out to sea. The exploitation of these areas for mineral resources such as oil is already a well-developed industry in most technologically advanced countries. A sketch of the mature Atlantic is shown in figure 2-5.

The gradual slope of the ocean floor as we move away from the mid-ocean ridge is due to two effects: the cooling and consequent shrinking of the new rock on the ocean floor, and the weight of the ocean water pushing down on the rock. The longer the rock is around, the cooler it will be, and the more it will be deformed by the weight of the water. If the ocean floor is indeed moving outward from the mid-ocean ridge, then the farther away the floor is from the ridge, the older it will be. We therefore would expect the Atlantic Ocean to be less deep at the center than it is at the edges of the continental shelves. In fact, the depth of the mid-Atlantic ridge is about 7,500 feet, whereas about six hundred miles away the floor is 50 million years old and has dropped to a depth of 15,000 feet.

A secondary effect of the increasing depth of the ocean away from the ridge is shown in the type of sediment that accumulates along the sea floor. We know that below the "snow line" marine skeletons and shells are simply reabsorbed back into the water. Below that line the top layers of sediment will be clay drifting down from the rivers, while above the line there will be a mixture of clay and calcium compounds.

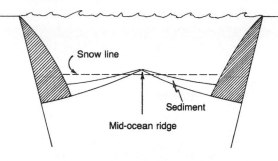

Figure 2-5.

There is an overwhelming amount of evidence for the theory of plate tectonics. Let me mention just one piece that fits in with the present discussion. In taking samples from the ocean bottom near the continental shelf, scientists often find a layer of calcium carbonate covered by a thick layer of clay. The calcium is at a depth of 18,000 to 20,000 feet—well below the "snow line." If the ocean floor had been at its present depth when this material fell to the bottom, no such layer could have accumulated, any more than real snow could survive for years in the desert. The only explanation for this widespread phenomenon is that when the calcium layer was laid down, the ocean floor was above the snow line. As the floor gradually moved down and away from the mid-ocean ridge, the calcium must have been covered over by layers of ordinary clay, which shielded the calcium from the water into which it would normally have dissolved. The calcium stayed in its protected hideaways for tens of millions of years, testimony to the forces that shape the world's oceans. The accumulation of hundreds of small bits of evidence like this is what has led to the revolution in the earth sciences embodied in plate tectonics.

Such are some of the profound insights that accompany the explanation of as simple a fact as the saltiness of the sea. If we consider the phenomenon in all of its ramifications, we are led to a view of the sea basically opposed to the view held throughout most of recorded history. Because the sea changes slowly, it has always been assumed to be static. We now know that it is far from static and unchanging. The Atlantic Ocean that I first saw as a boy on a family auto trip to Quebec is not the same ocean that my daughters delight in visiting on the Outer Banks of North Carolina. For one thing, it is wider now than it was then—more than a yard wider. The chemical factory in the ocean has been working away steadily all this time, and though thirty years is short compared to the residence time of calcium or sodium, there are other elements (such as aluminum) that have been completely processed and replaced in those same thirty years.

The Ionian Greek philosopher Heraclitus, in about 500 B.C., argued that everything in the world was in a state of flux and change. He put his idea tersely in the famous epigram: "No man steps into the same river twice." Our modern view of the dynamic ocean suggests that if we are willing to take a longer view of things, we can make a parallel statement: No child plays in the same ocean twice, either.

3

THE TIDES

"(Falstaff) parted even just between twelve and one, even at the turning o' the tide."

—WILLIAM SHAKESPEARE,
Henry V, Act II, Sc. III

No one who whas spent more than a few hours on a beach can fail to notice the dramatic changes in water level we call the tides. Many a novice camper has discovered them the hard way, bedding down yards from the water, only to be rudely awakened when the waves begin to soak through his sleeping bag. My own introduction to the tides was somewhat less startling, though equally effective. A group of students on spring vacation had stopped at a Youth Hostel on the coast of Portugal to enjoy a particularly fine afternoon. To while away the time, a friend and I started to build a large and complex sandcastle. Since he was from Texas and I from Illinois, neither of us paid much attention to the position of the castle relative to the water. Our first intimation of

23

approaching disaster came when a particularly large wave brought some water up against the castle wall. The ten yards that had separated our castle from the sea had shrunk to a few feet. We built a small sea wall of sand and proceeded unconcerned until another, larger wave intruded on our construction. As the afternoon wore on, the focus of our labors shifted from the castle itself to a series of barriers to protect it from the ocean. Holding back the tide became something of an obsession, and we hauled driftwood logs and large rocks to buttress our increasingly battered ramparts. When all else failed to save our handiwork, we hurled ourselves into the waves, all to no avail. Despite our efforts, we eventually saw our work vanish under the advancing water. We left the beach that day with a new appreciation of the forces of nature.

According to legend, there was at least one precedent for our attempt to stop the tides that April afternoon. King Canute the great, the eleventh-century ruler of most of England and Scandinavia, was supposed to have had his throne carried down to a beach, where he ordered the tide to stop. "Thus far shall you go, and no further," one author has him say. When the waves washed over his throne anyway, he used this as an object lesson to his courtiers on the limitations of power.

There are other references to the tides in folklore and ancient writings, though fewer than you might expect. Compared to other natural phenomena such as the phases of the moon, tides seemed to get scant attention in ancient times. We find them described by Herodotus in the fifth century B.C., but they receive little notice from other classical authors. That the Greeks, a seafaring people, spoke of the tides so seldom is usually attributed to the fact that tides in the Mediterranean basin are not very big—the difference between water levels at high and low tides is typically a matter of only six to twelve inches.

Perhaps the most widespread bit of folklore about tides concerns the supposed relation between death and the turning of the tide. A number of mediterranean writers are supposed to have believed that people could only die when the tide was going out. This belief was widespread in folklore along the Atlantic coast of Europe from Spain to Scandinavia, and survived in England well into the nineteenth century. Shakespeare has Falstaff die at the turning of the tide, and Dickens has Mr. Peggotty say that people can't die "except when the tide's pretty well nigh out." This belief in the tides as the mover of souls appears in a number of other places, including among Pacific coast Indians in North America.

But if folklore on this topic is relatively scarce, most of what there is does agree on one point: the tides are in some way associated with the moon. For example, among the Tlingit Indians of the Pacific North-west, the tides result from a battle between the moon (portrayed as an old woman) and the raven, who wins a narrow strip of land back from the sea in order to feed his brother animals. It wasn't until the development of the theory of universal gravitation by Isaac Newton that this idea of a vague connection between the moon and the tides could be replaced with a comprehensive tidal theory.

The phenomena any theory of the tides must explain are the following:

- In most places, there are two high tides and two low tides a day.
- High tides generally occur when the moon is on the horizon.
- Tides are highest during the new and full moon, lowest midway between these points. The high tides at new and full moon are called spring tides, the lower ones at the first and third quarter moon neap tides.
- The range of tides (that is, the difference in water level between high and low tide) is generally 3 to 10 feet, but can be much higher or lower in some locations.

We all learned in grade school that tides are caused by the gravitational pull of the moon on the oceans, and at first glance that seems like a reasonable point of view. The moon pulls the water up, so we observe a rise in sea level. But if that is what causes tides, shouldn't high tide occur when the moon is directly overhead? And why should there be two tides per day? The moon, after all, is directly overhead only once a day.

It is clear that the first, simple explanation of the tides has to be modified. Let me introduce some important ideas by considering the example below. If we ignore the fact that the earth and the moon are actually moving in space, then the effect of the moon on the earth's oceans is easy to predict. The force of gravity has the property that it decreases in strength with distance—the farther we are from a massive object like the earth or the moon, the weaker the force becomes. This effect is not confined to astronomical bodies; if you climb a flight of stairs, your weight will decrease by a thousandth of an ounce or so because you have increased the distance between yourself and the center of the earth. This feature of the gravitational force implies that the part of the ocean nearest the moon will be more strongly attracted than the part farthest away. The result of this disparity in attractions will be that the earth's oceans will arrange themselves as shown in figure 3-1, with the water deeper under the moon than it is on the opposite side of the earth.

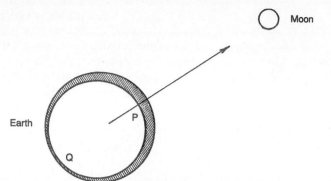

Figure 3-1.

The excess of water is called a tidal bulge. An observer standing at a point like the one labeled P would note that the sea level was high; he would also note that the moon was directly above his head. He would, therefore, say that high tide occurred when the moon was at its zenith.

Now let's bring the rotation of the earth into our thinking. Twelve hours after our observer was at point P, with the moon directly overhead, he will be at the point now occupied by an observer at Q on the opposite side of the earth. The ocean water rotates with the earth, of course, but provided the earth rotates slowly enough, the water will adjust so that the shape of the tidal bulge stays the same, with the water under the moon bulging toward it, even though the individual bits of water that compose the bulge are different. Someone standing on the earth and therefore rotating with it will see the water rise and fall once during each twenty-four hour day, a phenomenon very much like the tides, except for the fact that there are two high tides each day, not one.

This way of looking at ocean tides suggests a useful, though fictitious, picture of things. Suppose, for the sake of argument, that we ignore the fact that the oceans rotate with the earth. We can then think of the tides as follows: the gravitational attraction of the moon raises a tidal bulge as shown in the figure. The earth then rotates *underneath* the bulge, so that an observer rotating with the earth is carried alternately to regions of high and low water level. He will therefore see the water level rise and fall, and will interpret this as tides. Although this way of looking at tides is plainly not correct—the ocean does in fact rotate with the earth—the picture of the earth rotating underneath a tidal bulge is so easy to grasp that it will be convenient to use it as we pursue our discussion.

Having given a precise account of the simplest possible explanation of the tides, we now have to admit that the explanation must be wrong. It predicts one tide a day and it predicts that high tide will occur when the moon is overhead. Neither of these predictions agrees with what we actually find.

The two failures of this elementary view of the tides can be ascribed to different sources. The presence of a tide which crests twice a day—the so-called diurnal tide—stems from the fact that the earth-moon system is more complex than we have suggested. In particular, it is not true that the moon moves in orbit around a stationary earth. In point of fact, both the earth and the moon move in orbits around a point between them called the center of mass. Similarly, the failure of the model to explain the appearance of high tides when the moon is on the horizon is due to the fact that we have treated the movement of water in much too cavalier a fashion. The seemingly paradoxical relation between the position of the moon and the behavior of the tide can be explained only by a more complete understanding of the way that water behaves in large, relatively shallow basins like the oceans.

Imagine putting a bowling ball on one end of a yardstick and a billiard ball at the other. If you move your hand underneath the stick there will be one point where the two weights will balance, as shown in figure 3-2. This point is known as the center of mass of the system. If we

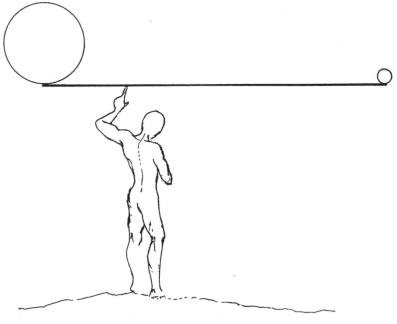

Figure 3-2.

support the stick as shown and push on the weights, then the entire arrangement will rotate about the center of mass.

The earth-moon system behaves in a similar way. The earth is eighty times as massive as the moon, so the center of mass of the system is only 1/80th of the distance from the earth's center to the moon—a point that lies about a thousand miles beneath the earth's surface. Over a period of one lunar month, the earth actually moves so that its center describes a circle around the center of mass, as shown in figure 3-3.

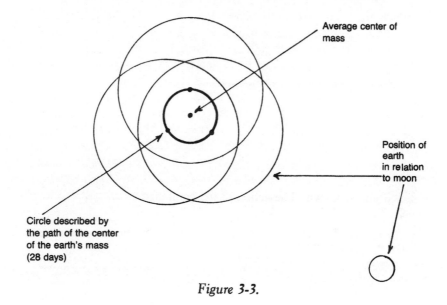

Average center of mass

Position of earth in relation to moon

Circle described by the path of the center of the earth's mass (28 days)

Figure 3-3.

Whenever a body moves in a circle, centrifugal force comes into play. You have undoubtedly experienced this force—it's what pushes you against the car door when you drive around a corner too fast. At any point during the lunar month, every object on the surface of the earth feels this kind of force pushing outward, as shown in figure 3-4. We don't notice this force very much because it isn't very large compared to other forces in the system. Nevertheless, it is present and is a part of the constant interplay of forces to which the motion of the earth and the force of gravity subject us. The important thing about this force as regards the tides is that it is the same everywhere on the earth; it acts in exactly the same way on water at point A, underneath the moon, as it does on point B, on the opposite side of the earth.

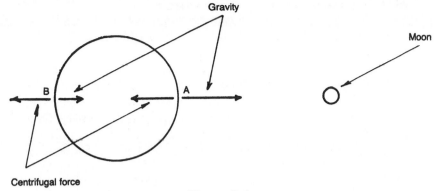

Figure 3-4.

The moon's gravity, however, does not act in this way. Because point B is farther from the moon than point A, the gravitational force on a stretch of water at B is somewhat less than it is when that same stretch of water is at A. The laws of physics tell us that the two forces—centrifugal and gravitational—must be equal in magnitude at the center of the earth. It follows, therefore, that at point A the gravitational force will exceed the centrifugal, while at point B the reverse will be true. Thus, the net force on the water at A and at B will be in opposite directions, as shown. The action of the combined forces will produce a tidal bulge of the type shown in figure 3-5. It will have two bulges 180° apart. If, as suggested above, we think of the earth rotating under this sort of tidal bulge, it is easy to see that we will observe two high tides each day. In other words, the appearance of a diurnal tide on the earth is a result of the fact that the earth moves in response to the gravitational pull of the moon.

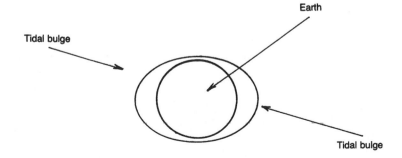

Figure 3-5.

One way to think of the production of the double tidal bulge is to note that the moon pulls the water away from the earth on one side, and the earth away from the water on the other. Both effects lead to a raising of the apparent water level. The diurnal tide is a result of the fact that the force of gravity acts on everything: earth and water alike.

What about the fact that the tides seem to occur at the wrong time, high tides appearing when the moon is on the horizon instead of when it is overhead? The resolution of this difficulty requires that we think about the oceans themselves. The large ocean basins of the earth are typically thousands of miles across, but the average depth of the ocean is only a few miles. The oceans should therefore be thought of as large shallow pans full of water. You can even make yourself a miniature "ocean" by putting a little water into a large baking pan. If you jostle the side of the pan, you will see waves moving across the surface of the water. This corresponds to the earthquake-generated "tidal" waves, or tsunami, that appear in the ocean. They are not related to the moon, and so do not represent true tides.

If, on the other hand, you rock the pan gently and rhythmically, the water will slosh back and forth. Someone standing on the side of the pan, as shown in figure 3-6 would see the water alternately rise and fall, for all the world like a tide. In fact, if you think of the pan as the ocean basin, the water as the ocean, and your hand as representing the force exerted on the ocean by the moon, you have a pretty good tabletop analogy to the tides themselves.

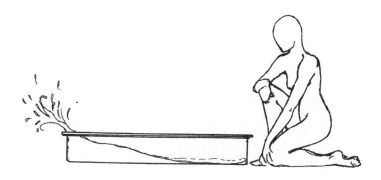

Figure 3-6.

You can imagine a very large baking pan wrapped around the earth's equator, as shown in figure 3-7. If we allow water to flow freely around the pan, we have a good representation of one of the first historical theories of the tides. We picture the tides as being caused by the motion of water moving around the earth in channels such as those enclosed by our baking pan. If we disturb the water in one of these channels (for example, by tapping the sides of the pan), a wave of water will move around the earth. Once the wave starts out, it will move with a speed which depends on the depth of water in the canal, but which will not depend on the way the wave is created. In general, the deeper the water, the faster the wave. An observer standing at a point along the channel will see the water level rise as the wave crest moves by, and then fall. In short, a wave moving in the channel will have the same sort of observable effects as a tide.

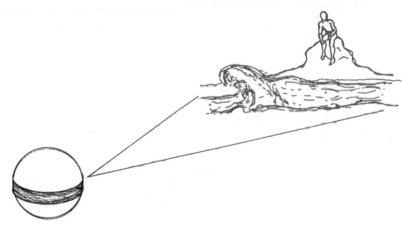

Figure 3-7.

If we now think of disturbing the water in the channel by means of the moon's gravitational force rather than by tapping the pan, we can convert this picture into an explanation of the tides. As we have seen, the moon raises a tidal bulge on the earth, and this tidal bulge has two lobes. If the raised part of the bulge is to stay directly beneath the moon, then the wave associated with each part of the bulge would have to move halfway around the world in about twelve hours—a speed exceeding a thousand miles per hour at the equator. In point of fact, a wave in an ocean three miles deep can travel at a rate of only five hundred miles per hour or so. So even though the moon's gravity causes the deformation that leads to the existence of the wave, the ocean itself is built in such a way that the wave, once created, cannot keep up with the moon. The moon actually races ahead of the tidal bulge. When all the mathematics is done, it turns out that the wave associated with the tidal bulge will always lag 90° behind the position of the moon itself. This means that when the moon is directly overhead, the tidal bulge, trying to catch up, will still be a quarter of the way around the world, as shown in figure 3-8. Thus, when the moon is directly overhead we see the lowest part of the wave (low tide), while when the moon is rising or setting we will see the crest of the wave (high tide). If the oceans were sixty miles deep instead of only three or four, then the tides would be direct and high tide would occur when the moon was overhead.

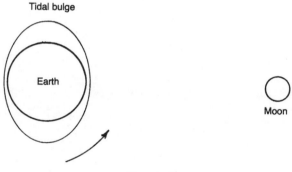

Figure 3-8.

The third point on our list of tidal phenomena—the variability of the tides over a lunar month—cannot be explained in terms of the earth and moon alone. It should be clear from our discussion so far that the gravitating body which raises the tides need not be the moon. Any body which exerts a gravitational force on the oceans will do the same. In principle, the planet Mars could raise tides on the oceans, but in practice Mars is too small and too far away to have any discernible effect. Only the sun, because of its huge mass, makes any difference as far as the tides are concerned. Like the moon, it raises a two-lobed tidal bulge that lags behind as it tries to follow the track of the sun across the sky. There are two situations, shown in figure 3-9, where the effects of the sun and moon will reinforce each other. The case shown at bottom, with the sun and moon lined up on the same side of the earth, is obvious. This is the configuration for a new moon, since none of the light reflected from the face of the moon will find its way to the earth. The second situation, shown at top, is not so obvious. At first glance, it might appear that when the sun and moon are on opposite sides of the earth, their effects would tend to cancel one another out. We must remember, however, that the sun and the moon each produce a tidal bulge on both the near and the far side of the earth. That means that in the configuration shown at the top—a configuration corresponding to a full moon—the tidal bulge raised by the sun on the far side of the earth will reinforce the bulge raised by the moon immediately underneath itself. At both new and full moon, then, we expect the tides to be highest, and this is the explanation of the appearance of spring and neap tides.

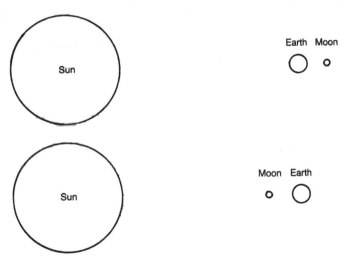

Figure 3-9.

Up to this point we have dealt with a rather idealized picture of the tides. There is nothing in what we've learned so far to explain the enormous differences in the tides seen in different parts of the earth. Why, for example, are the tides in the Mediterranean so small, while in the Bay of Fundy in Newfoundland the range can be fifty feet? Why is it that in some areas of the South China Sea there is only one tide a day instead of two? Why is that in Tahiti the tides are not correlated with the motion of the moon at all, but occur regularly at noon and midnight?

To understand these facts, and to get a glimpse into the difficulties that can arise when we try to apply a beautifully simple theory to the real world, we can go back to our analogy between the tides and an ordinary baking pan with a layer of water in it. We learned that an observer on the side of such a pan would interpret the regular sloshing of the water as a sort of tide. But now suppose that instead of a simple pan full of water, we include some barriers (bricks, for example that could impede the flow of the liquid. One such situation is shown in figure 3-10. It's not hard to see that if we tilted the pan by lifting the right-hand side, it would take the water a long time to flow into the basin formed by the bricks. An observer inside the basin, therefore, would see "high tide" occurring later than an observer outside. Similarly, it would take the basin longer to empty when the pan is tilted the other way. From this we conclude that the tides observed at a particular

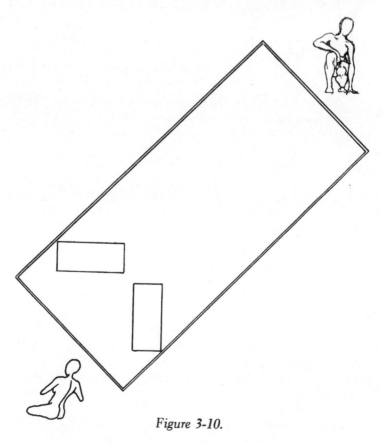

Figure 3-10.

spot on the earth will depend not only on the dynamics of the earth-moon system and the general structure of the oceans, but also on the details of the obstructions placed in the path of tidal motion.

A glance at a map or globe shows that, except for a narrow band of open water north of Antarctica, there is no place on the earth where tidal bulges can proceed unimpeded. Hence, the shape of the ocean basins and the continents that bound them must be taken into account if we want to make accurate predictions of the tides at any particular location. The irregularities in the shape of the land and the ocean bottom are responsible for the high tides in the Bay of Fundy, the daily tides in the China Sea, and the "solar" tides in Tahiti, as well as for a variety of other irregularities in other tides around the world. In general, a realistic calculation of tides is a grubby, messy job.

This illustrates an important but unappreciated aspect of the scientific enterprise. It is relatively simple to explain the general principles that govern a given physical phenomenon; gravity, centrifugal force, and the behavior of water waves give us a good general description of the tides. But to apply these principles in a specific practical situation may be far from simple. Physicists and others in basic research, trained to ignore details and penetrate to the core of a problem, often fall into the trap of underestimating the importance of those details. This aspect of the physicist's way of thought is illustrated by a joke about a technical adviser to the Navy antisubmarine program during World War II. He suggested that it would be easy to detect enemy submarines—all one had to do was raise the temperature of the Atlantic to boiling point and deal with the subs as they surfaced. When asked how to boil the ocean, he grew haughty. "That," he said, "is merely a technical detail."

As it happens, there is a rather interesting history connected with the complexity of the tides in real life. Their accurate prediction has obvious commercial advantages. In the days of sail, for example, a ship could slip its moorings and be carried out with the ebb tide, regardless of the strength and direction of the wind. In most of the major ports of Europe, private companies grew up whose main stock in trade was carefully guarded formulae for producing tide tables. These formulae were based on experience and tide records, and were probably analagous to the secret formula used by the Old Farmer's Almanac to predict the weather. Fortunately, tides are more regular and predictable than the weather, and by the nineteenth century the private companies were doing quite well in accuracy and, one supposes, in profits as well.

While these formulae were being developed for specific harbors in

Europe through the accumulation of experience and empirical "rule of thumb" calculations, the general principles governing tides were being studied by scientists. The theory of the tides we've been describing was first published by the French mathematical physicist Pierre-Simon Laplace in the late eighteenth century. It supplies an understanding of the tides on the basis of fundamental principles, but it ignores many of the details that have to be taken into account if we want useful predictions of the tides at a given spot. This may seem like a "mere technical detail" in the broad sweep of science, but it would not seem insignificant to a ship's captain who had to have his ship loaded by the next tide. In the early nineteenth century two Englishmen, William Whewell and John Lubbock, tried to use Laplace's new theory to predict the tides in the Port of London. The established companies did not take kindly to this sort of competition, of course. But their effort was advanced considerably by the general awareness that the theoretical tide tables contained a number of errors, all of which were pointed out by the entrepreneurs. It took some time for the tables generated by physicists to have an accuracy comparable to those produced by the secret "non-scientific" formulae. A number of prominent scientists, including Lord Kelvin, the founder of thermodynamics, contributed to making tide calculations reliable. Even today, predicting worldwide tides strains the capabilities of the biggest computers.

I think of the intrepid pioneers of tide prediction in much the same way as I think of meteorologists who try to predict the weather today. In both cases, the general principles governing the phenomenon are well understood, but so many details have to be included to make the calculations reliable that one is often defeated by sheer complexity. The National Weather Service, for example, still has only a sixty percent success on its short-term forecasts, and considerably less on the long-term ones.

It must be very discouraging to work hard, use the best of modern science, employ the best and fastest state-of-the art computers, and still not produce much better predictions than Aunt Bessy's rheumatism.

4

THE SEARCH
FOR
PLANET X

"By the work one knows the workman."

—FABLES OF LA FONTAINE

Although you may not have thought of tides in precisely this way, they are one of the few things on earth that bear witness to the effects of extraterrestrial influences on our lives. If the earth were continually blanketed in thick clouds and we could never see the heavens, we could still infer the existence of the moon from the tides.

We can make this statement thanks to Newton's Law of Universal Gravitation, which states that there will be a gravitational attraction between any two bodies in the universe. The larger the bodies, the stronger the force, but it becomes weaker as the separation between the

37

bodies grows. For reference, physicists write Newton's law in the following way:

$$F = \frac{Gm_1m_2}{d^2}$$

In this shorthand notation, m_1 and m_2 stand for the masses of the two objects under consideration, d is the distance between them, F is the force of gravity acting on each one, and G is a number known as the gravitational constant. In a system of units where mass is measured in kilograms and distance in meters, G has the numerical value of 6.67×10^{-11}.*

The important thing about this law is not its form, however, but the fact that it tells us that *every* object in the universe exerts a force on *every* other object. As you sit reading this book, there is a gravitational force pulling you toward the center of the earth, a force with which we are all quite familiar and which we call weight. At the same time, everything else in the universe is exerting a force on you as well. You are, for example, being pulled up toward the moon, out toward Mars, and even (very slightly) in the direction of the Andromeda galaxy. Most astronomical objects are so far away, of course, that the force they exert is unmeasurable. Nevertheless, it is present in principle.

Newton's law speaks of universal gravitation—there is no restriction to what can be considered an astronomical body. The moon does indeed exert a gravitational force on you, but so does this book. The sizes of the forces exerted on a 220-pound person by various objects are given below.

object exerting force (and mass in kg)	distance (in meters)	force (in pounds)
earth (6×10^{24})	6.4×10^6	220
sun (2.10^{30})	1.5×10^{11}	.13
Mars (6×10^{23})	7.7×10^{10}	1.5×10^{-7}
another person (100)	1	1.5×10^{-7}
book (.5)	.5	3×10^{-9}
nearby star (2×10^{30})	10 light years †	3×10^{-13}
nearby galaxy (10^{40})	2×10^8 light years	4×10^{-18}

* In this system, force is measured in the somewhat obscure unit called the newton. To convert to the more familiar unit of pound, just multiply the result of your calculation by 0.225.

† 1 light year = 9.4×10^{15} meters

From the table, we see that large nearby bodies exert the largest force, but that, at least in principle, the effects of more distant masses ought to be detectable with sufficiently precise instruments. There is, of course, a corollary to this statement. If the Andromeda galaxy exerts a force on you, then according to Newton's law, you also exert a force on it. Remember that the next time you cross the room, you are not only moving yourself from one place to another, but exerting a force (albeit a very small one) on every other body in the universe.

This property of gravitation can be used to detect distant masses, even if they are not seen directly. One application of the principle was the discovery of the outermost planets of the solar system. Another application, one that you can make, is to measure the mass of the moon with a ruler and a length of garden hose next time you go to the beach.

In the last chapter we saw how the combination of the moon's gravitational attraction and the centrifugal force associated with the movement of the earth in response to the moon's gravity produces the tides. Since both of these effects depend on the moon's mass, it should come as no surprise that the range of the tides (the difference between water level at high and low tides) should depend on the mass of the moon. The details of this dependence, as we saw, are likely to be very complicated, but the general features should be easy to understand. If we assume that the earth is covered with water, grant that the resulting ocean is deeper at the equator than at the poles, and take the ocean's average depth to be 15,000 feet, we can get a fairly simple expression for the tidal range. If we observe the tides at a latitude L, then the range of the tide will be

$$R = 60h \left(\frac{T_E}{T_m}\right)^2 \left(\frac{M_m}{M_E}\right)^2 \cos^2 L$$

where T_E is the time it takes the earth to rotate once on its axis (1 day), T_m is the time it takes the moon to complete its orbit (29.5 days), h is the average ocean depth, M_m the mass of the moon, and M_E the mass of the earth.

This result, like those discussed in the last chapter, is a gross oversimplification of the true picture of the tides. Nevertheless, the earth as we have imagined it so as to arrive at this expression is not very different from reality. The oceans *are* deeper near the equator than the poles, the surface of the earth *is* mostly water, and the oceans *do* have an average depth of around 15,000 feet. Therefore, we can hope that conclusions drawn from this equation will have some resemblance to the real world.

The mass of the earth is 6×10^{24} kg. The latitude, L, varies from one point to another, of course, but on the Outer Banks of North Carolina, where I did the measurement discussed below, it is 36°. The cosine of 36° is .81, so in North Carolina the range of the tide should be, assuming an average ocean depth of 15,000 feet,

$$R = 608 \left(\frac{M_m}{M_E}\right)^2$$

Values of $\cos^2 L$ to put in the formula for other locations are given in the table below. Since we know the mass of the earth, it follows that all we have to do is measure the difference between high and low tide and we'll have the mass of the moon. This is where the garden hose and the yardstick come in.

Latitudes

location	latitude	$\cos^2 L$
Brunswick, ME, or Eugene, OR	44°	.52
Philadelphia, PA, or Mendocino, CA	40°	.59
Kitty Hawk, NC, or Monterey, CA	36°	.66
Savannah, GA, or San Diego, CA	32°	.72
Miami, FL	26°	.81

The first step is to find out when the high and low tides occur on your beach. You can either do this by keeping watch yourself, in which case you may have to observe for a few days before you start the actual measurement, or you can get the information from local sources. Tide tables are usually posted in parks and harbors, and are invariably published in local newspapers. The first step in the experiment is simple: Go to the beach and mark the high tide line. I just drove a stake into the sand at a point I judged to be a typical high water mark, but any way of recording the spot will do. About six hours later, at low tide, you take your garden hose and yardstick back to the beach. Although it's not essential, your job will be easier if you take along a helper.

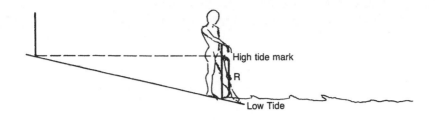

High tide mark

R

Low Tide

Figure 4-1.

When you get back to the beach, you will find that the water has receded a long way from the mark you left at high tide. Your job is to find the vertical distance (shown as R in figure 4-1) between the high and the low tide marks. Take the hose into the ocean and fill it with water. (The easiest way to do this is to uncoil it and hold it under the surface for a while.) Then, with you and your helper holding your thumbs over the ends of the hose to keep the water from running out, you move back up the beach. Your helper takes his or her end of the hose to the high tide mark, holding it at ground level at that point. You, in the meantime, take your end down to the water's edge and pick a spot that you will designate as the low tide mark. At this point you put a long thin stake into the sand.

You are now ready to begin the actual measurement. The object is to adjust your end of the hose until it is at the same height as your partner's. You will know you have achieved this goal if you can both take your thumbs off the ends of the hose and have no water flow out of either side. Actually achieving this state of affairs requires some trial and error, and you needn't get a perfect level to estimate the mass of the moon. Here's a technique that I found to work well enough: I raised my end of the hose until it was roughly level, marking the position on the pole. Then, on signal, my partner and I each took our thumbs off of the hose end for an instant and then quickly covered them again. The brief release was enough to tell whether water was running out at her end or mine. If it was running out the other end, then mine was too high. Accordingly, I lowered my end a bit and repeated the operation. We kept this up until the water ran out at my end. At this point we had bracketed the level point. We then replaced the lost water (from a water jug) and refined the measurement. On that particular day in North Carolina, we found the range of the tide to be about 3 feet 7 inches.

This result can now be analyzed. From the above equation, a 3'7"(3.6) range would imply that

$$\left(\frac{M_m}{M_e}\right)^2 = \frac{3.6}{608} = .0053$$

or, taking square roots, that

$$\frac{M_m}{M_e} = .07$$

This means that our estimate of the mass of the moon is
$$M_m = .07 \times 6 \times 10^{24} = 4.2 \times 10^{23} \text{ kg}$$

The actual mass of the moon is 7.4×10^{22} kg, which differs from our estimate by a factor of 6. Considering the crude model of the tides we used and the even cruder measuring equipment, this is an amazing agreement. Maybe you can do even better.

The important lesson from this exercise isn't the actual value we got for the lunar mass. In this age of satellites there are much more accurate determinations of this quantity. The important point is that we could detect and measure the gravitational influence of the moon without seeing that body at all. Indeed, there is a fifty-fifty chance that the moon will not even be above the horizon when you make your measurement—it wasn't when I made mine. An object need not be visible to us in order for us to detect its presence and measure its mass.

In our experiment on the beach, this lesson is brought home clearly but perhaps not as forcefully as it might be, for we know that the moon can be seen easily from the earth. The fact that it was below the horizon when my experiments were made doesn't really make the moon invisible in the usual sense. It turns out, however, that many significant objects in the solar system have announced their presence by gravity and not by light. During the course of a year, five different planets can be seen in the night sky: Venus and Mercury (as morning or evening stars), along with Mars, Jupiter, and Saturn. From ancient times, it was simply assumed that this was all there were out there, and astronomical debate dealt with the question of the structure and mechanics of the sun and the six planets. I strongly suspect that it never crossed anyone's mind that out beyond the orbit of Jupiter there might be other "wanderers," orbiting unknown to generations of astronomers.

On the night of March 13, 1781, William Herschel was observing the skies. Born in Germany, Herschel had emigrated to England, where his intense intellectual curiosity had taken him from his original profession of musician to amateur astronomer and telescope builder. Eventually, he started a survey of the heavens, using his own telescope, with the aim of cataloguing all the stars in order to determine the ultimate design of the cosmos. It was in pursuit of this goal that he was using his telescope that night. He saw a rather unusual fuzzy object in his lens and noted it as a "comet or nebula." Checking a few days later, he found that the object had moved relative to the stars and concluded it was a comet. It was as a comet that Herschel introduced his discovery to the European astronomical community. Only after several months of observation and calculation did it begin to dawn on people that Herschel's discovery was not just another comet but a new planet, so far from the sun that it could be seen only with a powerful telescope. Herschel's discovery was the first intimation that there was more to the solar system than could be seen with the unaided eye.

The name "Uranus" was chosen for the planet because it completed a mythological family in the sky. Uranus, god of the sky and husband of the earth, was the father of Saturn, the grandfather of Jupiter, and the great-grandfather of Mars, Venus, Mercury, and Apollo (the sun). I'm afraid that the days when scientists were sufficiently conversant with classical mythology to appreciate such fine points in naming new objects are long past. Witness the new convention in providing temporary names for new chemical elements, one of the latest of which is called *ununquandium* (that's 114 in Latin).

Throughout the remainder of the eighteenth century, scientists labored to incorporate Uranus into the accepted view of the solar system. We can get some idea of the complexity of this task if we think about the gravitational forces on the planet. The most important, of course, is that of the sun, since it is this force which holds Uranus in the solar system. If Uranus and the sun were the only bodies to be considered, the planet would move in an almost circular, slightly elliptical path, as shown on the left in figure 4-2. We know, however, that there are

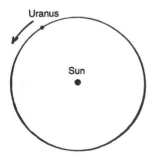
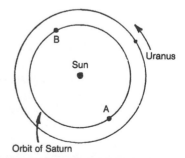

Figure 4-2.

other planets in the system. For the sake of illustration we show the orbit of Saturn on the right. (In this and subsequent drawings, the scale is distorted so that essential points can be emphasized.) When Saturn, the planet closest to Uranus, is at the point labeled A in its orbit, Saturn's gravitational attraction tends to pull Uranus back and slow it down. When Saturn is ahead at the point labeled B, this attraction tends to speed Uranus and the normal smooth motion of the planet is disturbed. As observed from the earth, Uranus first moves too slowly, then too quickly.

Saturn circles the sun every thirty years, while Uranus completes a circuit in a much more leisurely eighty-four years. Saturn always exerts a force on Uranus, but once every revolution this force is at its maximum because the two planets are as close together as they can get. It is during that period that the pertubation of the orbit of Uranus due to the presence of Saturn is greatest, hence it is then that the effect of Saturn on Uranus is easiest to observe from the earth.

In principle, every other planet in the solar system will produce the same sort of effect on Uranus, but most of them are too small and too far away to produce large pertubations. For all practical purposes, only Saturn and Jupiter, the largest planets, need be taken into account when we study the orbit of Uranus. Jupiter produces a slowdown and speedup every twelve years, which is the time it takes that planet to go around the sun.

For these reasons, the discovery of Uranus presented the astronomers of the 1780s with two problems. First, they had to establish the details of the planet's orbit; and second, they had to show that the orbit could be wholly explained by the workings of Newton's law.

Theoretically, all that is needed to carry out the first task is to watch the planet while it makes one circuit around the heavens. For Uranus, however, the eighty-four years required to do this would strain the patience of most astronomers; some substitute for this lifelong process had to be developed. Of the methods employed, two—one mathematical and one historical—were especially important.

The historical method goes by the name of "pre-discovery sighting" and depends on the fact that the planet may have been seen and recorded by astronomers who noted its position but, either through bad luck or oversight, failed to realize that what they were looking at was a new planet and not a faint star. In 1756, for example, the German astronomer Tobias Meyer recorded a faint star in a position where Uranus might have been expected to be at that time and in which no star could be seen in 1781. It was concluded that Mayer had actually seen Uranus and not realized the significance of what he saw. The 1756 point was therefore added to the post-discovery observations by those astronomers trying to trace the orbit of the new planet. Eventually, a search through the old records yielded pre-discovery sightings as far back as 1609.

This sort of data was extremely important to those attacking the orbit of Uranus from a mathematical point of view. The problem can be seen easily if we look at figure 4-3. In one year after discovery, the planet will have moved 1/84th of the way around its orbit—from the point labeled A to the point labeled B. With nearly every astronomer in the world following its progress, this part of the tiny piece of the orbit would be well charted. But many ellipses can be drawn through this region that are reasonable approximations to the true orbit between A and B, but which diverge widely from it at other points. A few of these are shown in the figure. It is impossible, on the basis of mathematics alone, to define the true orbit in this situation. If, however, we have a third point like the one labeled C, we can eliminate many of the false leads. Pre-discovery sightings supply these kinds of points.

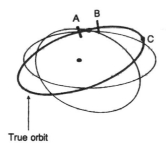

Figure 4-3.

During the remainder of the eighteenth century and well into the nineteenth, astronomers wrestled with the problems of understanding the motion of Uranus. A pattern soon emerged. No sooner would someone produce an orbit that took account of old sightings and matched the relatively precise post-discovery data than the observations of the planet would begin to disagree with the new predicted orbit. A number of well-known astronomers blotted their copybooks during this period. As the observations accumulated and the mathematical techniques increased in sophistication, the problem seemed to get worse.

presence of an advanced civilization. He wrote a number of popular books on the subject (including *Mars as the Abode of Life*), playing a role in early twentieth-century America somewhat analogous to that played by Carl Sagan today. But in the midst of the publicity surrounding the canals of Mars, Lowell directed and carried out a search for a new member of the solar system, a body he called Planet X.

In 1913, Lowell collected the best available data on the orbits of the outer planets and gave them to his computer. In those relatively simple days, the term "computer" was used to designate people who made a living by carrying out laborious calculations by hand (in fact, the obituary of the last computer from the Greenwich Observatory in London appeared in the papers only a few years ago). Lowell's computer, Elizabeth Williams, produced predictions for the orbit of Planet X that Lowell announced at a meeting of the American Academy of Arts and Science in 1915 under the title "Memoirs on a Trans-Neptunian Planet." Unfortunately, his death in 1916 and the legal problems associated with his will prevented the Observatory from mounting a major search for Planet X for some time.

In 1929, Clyde Tombaugh, a Kansas farm boy and skilled amateur astronomer, was hired by the Observatory to start a systematic search for the planet Lowell had predicted. The technique used was to make successive photographs of a particular region of the sky, then search for faint objects which had moved a small distance over the period of a week or so. (The requirement that the motion be small was needed to weed out the thousands of fast-moving asteroids that the telescope picked up.) In a series of photographs taken in January 1930, Tombaugh finally found what he was looking for: a small, faint point that had moved between the time of the two photographs. After a period of checking and calculation of orbits, the announcement went out—Planet X (or Pluto, as it was soon named) had been found. Once again, a distant body had announced its presence to us via its gravitational effects instead of visually. As the process of searching for pre-discovery sightings was carried out, made easier by the existence of large collections of photographic plates, a major irony emerged. In 1915, the year that Lowell published his memoir, the image of Pluto had been recorded on a plate taken at Flagstaff. This striking confirmation of his prediction lay hidden for fifteen years before being brought to light.

It would be pleasant if we could close this chapter neatly with an unequivocal statement that after the discovery of Pluto no more problems with planetary orbits remain, and that there are no more Planet

Xs out there. But the situation isn't quite that clear-cut.

In 1978, James Christy and Anthony Hewitt of the U.S. Naval Observatory obtained some very high quality photographs of Pluto and discovered that the outermost planet has a moon (subsequently christened Charon). By observing the revolution of the moon around the planet, an accurate determination of Pluto's mass can be made. The result: Pluto has a mass only .001 times that of the earth, or one-tenth that of the moon. Pluto's average distance from the sun is about forty times that of the earth, while the average distance of Uranus is about twenty. If Pluto is really as small as these measurements indicate, then the gravitational force exerted on Uranus by Pluto will be less than that exerted on Uranus by the moon! In other words, if there is really an unexplained perturbation in the motion of Uranus, Pluto will not make things right. The scientific consensus today is that if one does a careful analysis of the errors involved in the old measurements of the orbit of Neptune, there is no deviation to be explained.

Planet X, where are you?

5

THE BACK
OF THE MOON

When once around the earth hath run
The Lifesman knows the day is done

—PLAINSONG OF PERKINS, GENT.

Everyone who lived through the sixties will recall some particularly dramatic event associated with the Apollo space program. For most people, it was the moment Neil Armstrong became the first man to set foot on the surface of the moon. For me, though, an earlier event has pride of place. It was during the Apollo 8 mission in December 1968. There was live television coverage, of course, and while the spacecraft was still in earth orbit, we heard one of the astronauts announce laconically, "All systems go for TLI."

The announcer explained that TLI stood for trans-lunar injection. It took a while for this to sink in. Trans-*lunar* injection—my God, they were going to the moon! They would be further from earth than any human had ever been. They would be the first men to see the far side of the moon. I was stunned that, at the start of the greatest voyage ever undertaken, all they could say was "All systems go."

Since that time, in this marvelous book *The Right Stuff*, Tom Wolfe has explained the test pilot ethic that guided our astronauts, an ethic which includes the requirement that the most important statements be made as undramatically as possible, preferably in a southern drawl. But the fact remains that those astronauts, and a few of their breed that followed, are still the only human beings who have seen the far side of the moon. The rest of us must be content with photographs and the age-old face that has presented itself throughout the lifetime of the human race.

The constancy of the face the moon presents to us is so familiar a fact that we take it for granted; but if you think about it for a moment, you will realize that it is truly remarkable. In figure 5-1 we show the unseen side of the moon as a darkened hemisphere. The moon goes around the earth once a month, so if the moon didn't rotate about its own axis, we would have a situation like that shown on the left. If the familiar face of the moon were pointed at the earth (and hence visible) at point A, then two weeks later the moon would be at point B with the other side toward us. Since we never see the other side, we can conclude that the moon must indeed rotate around its axis, just as the earth does.

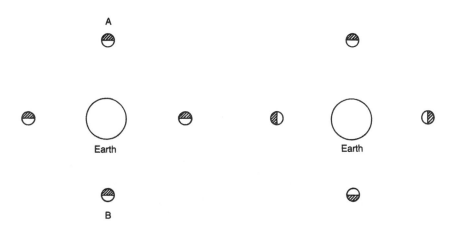

Figure 5-1.

A little thought will convince you that this rotation—the length of the lunar "day"—must be such that the moon turns once on its axis in exactly the same length of time that it takes to complete an orbit around the earth. This situation is shown on the right. It is, in fact, the only situation in which the moon would always present the same face to us.

But this is amazing. Of all the possible lengths for the lunar day, why should it be exactly as long as the lunar month? There is no obvious physical connection between the rotation of a body about its axis and the time required to complete an orbit. For example, the length of the earth's day is very different from the length of its year. Yet in the case of the moon, these two numbers are exactly equal. Can this be pure chance? The natural reaction is that it can't be—there must be some reason for this remarkable equality.

There is such a reason, and that reason has to do with tides. We are used to thinking of tides as the rise and fall of water in the ocean, but if you recall the origin of the tides in the interplay between the moon's gravitational attraction and centrifugal force, you can see that this is much too narrow a view of things. The forces responsible for tides will act on ocean water, raising tidal bulges as we discussed. They will also act on solid bodies, producing exactly the same sort of effect. A bit of solid material on land will experience exactly the same force as that which causes tides in the water. The solid land will not respond as much as the water, of course, but it will respond.

Figure 5-2.

We should expect, therefore, that a tidal bulge will be raised on land. The shape of the earth, for example, will be distorted by the tidal force into the familiar two-lobed shape shown in figure 5-2. This represents a compression of the solid earth itself, and would be present whether there were oceans or not. Someone standing on the earth, then, will see the land rise and fall twice a day, just as he sees the ocean rise and fall. The total movement of the land is so small—a matter of a few inches—that it is imperceptible to our sesnes, although it is clearly detected with delicate seismographic instruments. This sort of phenomenon is usually called an earth tide or land tide to distinguish it from the more familiar and more readily observable ocean tides.

If we shift our attention from the earth to the moon, the significance of land tides becomes apparent. There are no oceans on the moon, but there will be tides nevertheless. We should expect that the gravitational attraction of the earth would raise a tidal bulge on the moon exactly like the one shown in figure 5-2. Someone standing on the moon would see a tidal bulge go by twice during the lunar month, or about every two weeks.

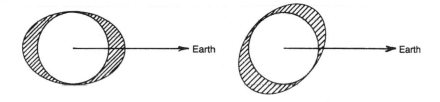

Figure 5-3.

If the moon were perfectly elastic—that is, if it were made of a material like rubber which could respond instantly to any forces applied to it— then the tidal bulge would be aligned so that one lobe was directly underneath the earth, the other directly opposite the earth. This situation is shown on the left in figure 5-3. In point of fact, no body in the universe is perfectly elastic. There are always imperfections and internal sources of friction that slow the response of a solid to an applied force, and this in turn guarantees that it will take a little time for the solid to adjust itself to changing external circumstances. On the moon, you can imagine bits of rock straining and grinding to form the tidal bulge (an exaggeration, of course, of what really happens), with a lag between the time when the force is largest and the time when the deformation of the solid is largest. You can think of this as being analogous to the fact that if you squeeze a rubber ball and then let it go suddenly, it takes a while for the ball to regain its shape.

During this time lag, while the rock is struggling to rise in response to the tidal force, the moon will continue to spin on its axis. If we assume for the sake of argument that the moon, like the earth, takes less time to turn on its axis than it takes to complete an orbit, then the result of the time lag will be a displacement of the tidal bulge. While the bulge is rising, it will actually be carried ahead of the line joining the centers of the earth and moon, a situation shown on the right in figure 5-3.

The fact that the moon is not a perfect solid—that there are sources of internal friction in its structure—leads directly, then, to a situation in which the tidal bulge raised by the earth is not placed symmetrically on the moon's surface. The earth's gravity does not act on a moon in the shape of a perfect sphere, but on a moon slightly distorted by an asymmetrical tidal bulge. An exaggerated view of such a moon is shown in figure 5-4. The earth's gravity affects every particle in the moon, of course, but throughout most of the moon this fact does not affect the rate of rotation. For example, there is a force on a particle at the position labeled A that would tend to turn the moon clockwise, but this tendency is exactly canceled by the force on a particle at a point B—a force that would tend to turn the moon counterclockwise. If the moon were exactly spherical, there would be a point B for every point A and the earth's gravity would produce no net rotation of the moon.

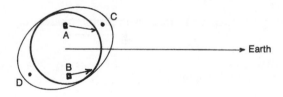

Figure 5-4.

When we get to the material in the tidal bulges, however, the situation changes. The points labeled C and D are typical of those that would have to be paired in this case. Once again, we see that the effect of the earth's gravity on C is to turn the moon clockwise, while its effect on D is to turn it counterclockwise. But C and D are not the same distance from the earth. This means that the force acting at C, because the distance to the earth is less, will be stronger than the force acting at D, with the result that the force tending to produce clockwise motion will overcome the force tending to produce the opposite. The net effect is that the earth's gravity, acting on the moon's tidal bulge, will tend to slow down the moon's rotation.

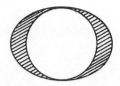

Figure 5-5.

But if the moon rotates more slowly, then the tidal bulge will not be carried as far forward as it is rising, and the asymmetry of the bulge will decrease. The shape of the moon, in other words, will go through a progression like that pictured in figure 5-5. If the moon is spinning too fast, so that the tidal bulge is carried forward, then the action of the force which produces those tides in the first place will slow that spin down, so that the tidal bulge is carried forward less and less. The end result of this process, of course, is that the moon will eventually turn at just the proper rate so that the tidal bulge is located just underneath the earth. Astronomers say that the moon is "despun," and you should be able to convince yourself that in this configuration the tidal forces produce no further effect on the rotation rate.

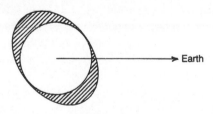

Figure 5-6.

Suppose we had assumed that the lunar day was much longer than the lunar month, rather than vice versa. What would the effect of the tides be in that case? An observer on the moon would then see the earth pass swiftly overhead. By the time the solid ground had had time to rise up in response to the tidal force, the earth would have moved far ahead. The situation would be like that in figure 5-6. The earth would run ahead of the tidal bulge, and an argument exactly like the one given above would show that the earth's gravitational forces would speed up the spin of the moon until the tidal bulge was directly under the earth.

No matter what spin the moon starts with, then, it must always end up in the same way. Its spin must adjust itself so that the tidal bulge follows the earth. The lunar "day," in other words, must always be *exactly* as long as the lunar month. Any deviation will result in a gravitational effect which will push the moon back to the equilibrium position.

So we never see the other side of the moon because long ago the moon was despun by the tidal forces created by the earth. This would be a fairly quick process as geological time is reckoned. For example, if the moon were in its present position, the entire despinning process would take only about 10 million years—less than 0.1 percent of the lifetime of the earth-moon system. Tidal despinning is actually seen in a number of places in the solar system. The inner satellites of both Jupiter and Saturn, the largest planets, have days and months that are exactly equal. Tidal effects caused by the sun's gravity have also had an influence on Venus and Mercury. The other planets are too far from the sun to be despun to any appreciable degree.

If the gravitational field of the earth can have an effect on the rate of rotation of the moon, then it is reasonable to expect that the gravitational field of the moon should have a similar effect on the earth.

We can think of two mechanisms by which such an effect might manifest itself: earth tides (appropriately named in this case) and ocean tides. Since the earth rotates once a day, while the moon takes a month to complete its orbit, the earth does not always keep the same face toward the moon; it has not yet been despun. The earth's day is shorter than its month, so the conditions are appropriate for the bulge associated with the earth tides to be pulled slightly ahead of the line joining the centers of the earth and the moon. The ocean tides are inverted, as we saw in chapter 1, because of the shallowness of the ocean basins. The effects of friction on the ocean tidal bulge cannot be discerned with a casual glance, but have to be worked out mathematically. It turns out that the effect of friction on tides in a shallow ocean is to make the bulge lag slightly behind its expected position. The net effect of friction on the two tides that exist on the earth is shown in figure 5-7.

Both the tidal bulges shown have the effect of slowing down the rotation of the earth. That the earth is not totally despun like the moon

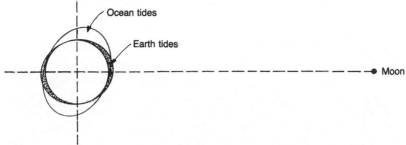

Figure 5-7.

has to do with its larger mass—for a given tidal force, a larger object will take longer to slow down than a smaller one. Nevertheless, a change in the rate of spin of the earth would be a dramatic effect, for it would change the length of the day. If the earth is being despun, then we expect that the length of the day must be increasing now, and must have been less than twenty-four hours in the distant past.

Before examining the evidence for the length of change in the day, let us note one more consequence of asymmetric tidal bulges in the earth-moon system. Up to now we have considered only the effects of the earth's gravity on the moon's tidal bulge, and vice versa. But tidal bulges, once formed, can also have an effect on the orbit of the moon around the earth. For simplicity, let's talk only about the effect of ocean tides on the moon. In figure 5-8, we show the earth-moon system with the tidal bulge as it exists, lagging slightly because of the effects of friction. The mass in the close bulge, labeled A, will exert a stronger attraction on the moon than the mass in the bulge labeled B. Since the near bulge is ahead of the moon, the effect of this gravitational force will be to accelerate the moon—to pull it forward in its orbit.

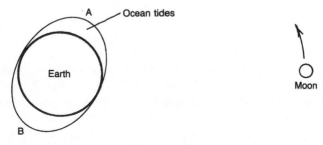

Figure 5-8.

This acceleration results in an increase in the moon's velocity, which, in turn, causes the moon to move farther away from the earth. This consequence arises through the action of centrifugal force; the faster the moon moves, the more it is thrown to the outside and the farther away from the earth its orbit moves. Because the new orbit is larger in circumference than the old, we have the somewhat paradoxical result that increasing the speed of the moon in orbit actually causes the length of the month to increase. Therefore, the orbit of the moon can be expected, over a long period of time, to describe a spiral moving to progressively farther distances from the earth, as shown in Figure 5-9.

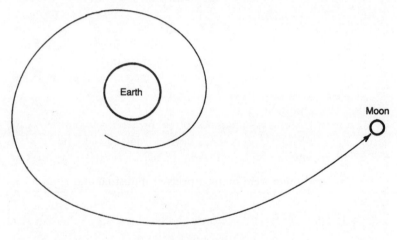

Figure 5-9.

Because the ultimate cause of this outward spiral is the presence of tidal friction in the oceans, and because the lag of the tidal bulge produces both a slowdown in the earth's rotation and an increase in the distance to the moon, physicists sometimes say that tidal friction provides a way for energy to be transferred from the earth's rotation to the orbital motion of the moon. This statement is true, but if I had started our discussion of tidal effects by putting it in that way, everyone (including me) would have been mystified. This is a good example of how jargon gets started in a science. The first time something is discovered, everyone goes in laborious detail through all of the steps that trace the connection between the beginning and the end. Later, scientists start throwing around statements like the one above, and only those willing to go through the process of rediscovering the step-by-step reasoning can understand what is being said. I suppose an anthropologist would recognize this pattern of behavior.

At any rate, the end result of all of this thinking about tidal effects in the earth-moon system is the following list of phenomena:

1. The earth's gravity acting on the moon's tidal bulge long ago despun the moon, so that it now always keeps the same face toward us.
2. The moon's gravity acting on the earth's tidal bulge is in the process of slowing down the earth's spin, thereby increasing the length of the day.
3. The earth's tidal bulge, acting on the moon in orbit, is causing the distance between the earth and the moon to increase.

This is an impressive list, particularly as we have seen how all these effects seem to follow from simple applications of Newton's law of gravitation. But how do we know that these predictions of the thoery are actually born out by observation? Except for item 1, there is no evidence readily available to our senses to confirm them.

As it turns out, the length of the day is a quantity of fundamental interest to astronomers. All units of time used to be defined in terms of the day, and there were many aspects of industrial and commercial life that depended on accurate measurements of time. Navigation, for example, depended crucially on the availability of accurate clocks, a fact which explains (in part) why one of the first major scientific establishments in this country was the Naval Observatory in Washington. Precision measurements with modern atomic clocks show that the day is indeed getting longer, at the rate of about two milliseconds per century. This means that a day in the twentieth century is two milliseconds longer than one in the nineteenth, so that this century will be a little over a minute longer than the nineteenth, which in turn was a minute longer than the eighteenth, and so on. There seems little doubt that in modern times, at least, the earth is being despun.

A somewhat longer historical perspective on this process can be gained by using an observation made by John Wells of Cornell University in 1962. Certain types of coral in the fossil record exhibit bands in their solid shell-like parts called epitheca. These bands represent annual growth increment, and as such can be thought of as similar to tree rings. Close studies have revealed that each yearly band was broken down into finer bands that seemed to correspond to monthly growth, and within these, still finer bands can be seen. These finer bands were taken to represent the daily growth of the coral shell. Wells counted the number of daily bands within the annual band for coral which had grown during the Devonian period—about 400 million years ago. He found that the average number of days in a Devonian year was 400, while in the Carboniferous period (roughly 300 million years ago) the number was 380. Simple arithmetic shows that the Devonian day must have been between 21 and 22 hours long, while the Carboniferous was around 22½. Both, of course, were shorter than the present day which, at 24 hours, occurs 365 times in a year. These numbers, with some uncertainty associated with distinguishing fine rings on old fossils, are precisely what would be predicted if the presently observed rate of despinning were to be extrapolated backward in time almost half a billion years.

The tides, then, play a role in a much bigger arena than human commerce and recreation. It appears that we cannot think about the tides without being led to contemplate their effects on the entire history of the earth-moon system. If the day is getting longer now, with the moon moving farther from the earth each year, it is only natural to ask whether we can use our knowledge of the effects of the tides to trace the history of the system back to its beginning and forward to its ultimate end. Since the extrapolation into the future is relatively straightforward and uncontroversial, we'll discuss that first.

The present tidal interaction which is despinning the earth and enlarging the lunar orbit will continue until the moon has receded to an orbit some sixty percent larger than the one it has at present. By the time this happens, the rotation of the earth will have slowed down to the point where the day is as long as the lunar month, and both will be about fifty-five days (where by "day" we mean twenty-four of our present hours). The tidal bulges of both bodies will stay aligned with the line joining their centers, as shown in figure 5-10. The moon is now moving away from the earth at the rate of about three centimeters per year, so it will take several billion years for the system to get into this final configuration. If we ignore the fact that the sun will turn into a red giant and swallow both the earth and the moon at about this time, we can, as an intellectual exercise, follow the earth-moon system even further into the future. The solar tides, although they have a smaller effect on the earth than those caused by the moon, will continue to operate. They will tend to slow the rotation of the earth. Thus, the earth's tidal bulge would lag behind the moon, and the moon would act to speed up the earth's rotation, overcoming the braking effect of the sun. The moon would also start to spiral back in toward the earth. (The argument here is just the reverse of the one we gave on p. 57.)

The whole business reminds me of a yo-yo.

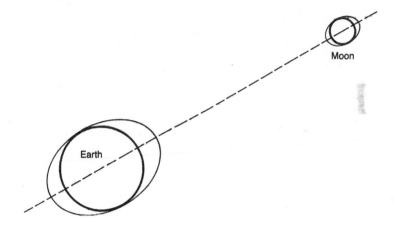

Figure 5-10.

If we use our knowledge of tidal effects to reconstruct the past instead of forecasting the future, the situation gets more interesting. Assuming that the angle by which the tidal bulge lags behind its 90° position is constant in time, then calculations yield the rather astonishing result that the moon and the earth were separated by no more than a few earth radii about two billion years ago! This is less than half of the 4.5 billion years that is usually accepted as the lifetime of the earth-moon system. This calculation, together with the finding of the Apollo missions that the surface of the moon has been in its present state for most of its lifetime, raises major difficulties linked with an old scientific question: Where did the moon come from?

In general, theories of the origin of the moon fall into three categories: fission, capture, and binary formation. The oldest (and least credible) of these is fission. It holds that the moon was once part of the earth and split off because the earth was rotating too fast. In some pre–continental drift versions, the Pacific basin was suggested as the "scar" marking the birth of the moon, although we now know that geological processes would have obliterated such a scar long ago. The main problem with this explanation of the moon's birth is that the earth's day would have been about five hours long when the moon split off. This rate of rotation is much too slow to tear the earth apart. To break up the planet, the day would have to be less than 2.6 hours.

On the other hand, the density of the moon (3.3 gm/cm³) isn't too different from that of the earth's crust. The main difference between the two bodies seems to be the large amount of iron in the earth's core. This element is not present in great quantities on the moon. It's hard to know what to make of such data. One of the original analyses of the moon rocks appearing in *Science* pointed out that the seismic properties of the rocks were quite similar in some respects to certain kinds of cheeses (Muenster and Cheddar). Had the authors been able to remove their tongues from their cheeks long enough, they would doubtless have pointed out that the fact that two materials have similar properties does not imply they were once part of the same body.

The second class of theories suggests that the moon was formed elsewhere and captured into the earth's orbit by a close encounter. Such theories explain the difference in composition of the two bodies, but have problems with the fact that when a body is captured, some energy must be dissipated in the system. For example, when a satellite is sent toward the moon, it will swing around and go on its way unless retro-rockets are fired to slow it down. Tidal friction is the easiest way to slow the moon so that it can be captured, but this would probably be a pretty disruptive process. Calculations show that the moon would have had to approach within 2.8 earth radii of the earth, a distance at which one has to worry about the moon being broken up by the earth's gravitational field. Furthermore, the moon would have to be moving at just the right velocity and in just the right direction for capture to occur at all. Scientists always feel uncomfortable when they have to explain things in terms of highly improbable events. Improbable and impossible are not synonymous, however, and there are still astronomers who favor the capture theory.

The binary formation theory can be thought of as a limiting case of the capture theory in that it envisions the moon and the earth forming in the same region of space, but as separate objects. This theory avoids the difficulties of the capture hypothesis since the earth and the moon would have been born moving at almost the same velocity. It is difficult, though, to understand how the iron content of the two bodies could be so different if they really came from the same raw materials. It would be like baking two pieces of dough from the same mix and having one piece wind up as a loaf of bread and the other as a cake. Nevertheless, some version of this theory seems to be held today by more astronomers than any other.

We should point out that the seriousness of the perceived problem would be considerably less if the rate of increase in the radius of the moon's orbit were less in the past than it is today. Data from fossil corals show no show no such trend in the past 500 million years, but that covers only about ten percent of the lifetime of the earth. We know that the shape of the ocean basins plays an important role in determining the behavior of the tides. We also know that the oceans have changed dramatically in recent geological times (see chapter 4). It is not impossible, therefore, that the shape of the tidal bulge could have been radically different in the distant past, so that the discrepancy between the calculated 2 billion year time to the moment of closest approach and the 4.5 billion year age of the moon could disappear. Greater discrepancies than this have vanished in the history of science. Recent calculations have added strength to this suggestion. They show that taking the changing rotation rate of the earth into account leads to lunar orbits only 60 percent smaller than their present size as far back as 4 billion years ago.

I do not find it at all unusual that scientists should still be debating the origins of the moon a decade after the manned landings of the Apollo program. It is the nature of scientists to be contentious, and no question that touches on so many different fields of expertise is going to be settled easily. What I find surprising is that the new geological and chemical data brought back from the moon have done so little to change the general features of the debate. The same three options—fission, capture, and binary formation—that were being debated before the program are being debated now. The extra data provided by the astronauts seems rather to have made it more difficult to concoct an acceptable solution. One astrophysicist, Peter Goldreich, summed up the situation by saying, "It is likely we shall still be speculating about the origin of the moon a century hence."

Note: Today, the reigning theory about the origin of the Moon is called the "Big Splash." According to the theory, which seems well verified, late in the Earth's formation the planet was struck by a Mars-sized object. The Moon was formed in orbit from debris thrown into space by the collisions. Since the debris came from the (relatively) lightweight mantle of the Earth, the difference in composition between the two bodies is explained.

6

MAKING WAVES

What goes up must come down.

—Anonymous

Growing up in Chicago, I found something about Lake Michigan very puzzling. When I went with my family to the Indiana Dunes, on the eastern shore of the lake, the waves rolled up on the shore, moving toward the east; but when we went to Wisconsin, on the western side of the lake, the waves also moved toward that shore, to the west. It seemed to me that if the waves always moved toward the shore, there must be a hole somewhere in the middle of the lake.

For all the familiarity that human beings have had with waves on water, it wasn't until well into the nineteenth century that a full theoretical understanding of the simplest waves was gained. Even today, with powerful computers and a century and a half of mathematical thought behind us, we are learning new things about waves in water, air, and solid materials. To give just one example, a few years ago I heard one of the world's leading theoreticians in the field of fluid motion describe some research he was conducting on the problem of what happens to the water in an enclosed harbor when a large wave such as a tsunami (tidal wave) arrives. This problem has obvious practical importance. I believe, in fact, that the research was being supported by the United States Navy. But I found as interesting as the clever work being described the fact that this seemingly simple problem hadn't been solved a century earlier. It is only slightly more difficult, after all, than describing how the water in your bathtub sloshes around after you put your foot down at one end.

From my point of view as a theoretical physicist, one of the great attractions of the field of fluid mechanics is the fact that there are many interesting and relatively simple problems like this waiting to be investigated. That they haven't been solved up to this point is sometimes a matter of oversight, sometimes a matter of unanticipated complexity. But whether we are thinking about a bathtub, a lake, or an ocean, sooner or later we find ourselves thinking about waves.

A wave is a strange thing. As you watch the waves coming in toward the beach, you are seeing something moving *on* the water, which, at the same time, is not primarily a movement *of* the water. You can convince yourself of this by picking up a piece of driftwood and throwing it out past the point where the waves are breaking. As each crest moves by, you will see the driftwood rise and fall. The floating wood is locked into the water around it and its motion is the same as that of the water. From this experiment, we conclude that the passage of the wave causes the water to move principally in an up and down direction. Yet we *see* the wave itself moving toward the shore. The wave moves in one direction while the water moves in another. Although a wave could obviously not exist if the water were not there, it is nevertheless something more than just moving water. It is an organized, collective motion of particles of water, a motion which behaves according to its own sets of laws.

Whenever a physicist wants to analyze motion of any sort, the first thing he thinks about is forces. What forces are acting in this situation, and with what effect? Two forces are in fact acting on every bit of water: the earth's gravity, which pulls the water downward; and the pressure of the water itself, a force representing the effect of collisions between water molecules. For a stretch of water at the surface, the effect of pressure is an upward force, a force we sometimes call buoyancy. In a completely undisturbed system, these two forces—one upward and one downward—balance each other exactly, producing a smooth, even surface.

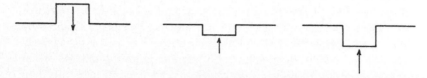

Figure 6-1.

If some disturbance acts on the surface, this simple situation changes. By way of illustration, suppose that the surface of the water is displaced in the highly unlikely configuration shown on the left in figure 6-1. In the portion of the water which has been raised above the normal level, the pressure force is less than the force of gravity, so the net motion of the raised portion will be downard. It will fall, just as this book would fall if you dropped it. As the water falls back toward equilibrium, it gathers speed; when it reaches that level, its velocity will prevent it from stopping and it will overshoot the level of equilibrium, giving us a situation like that shown in the middle of the figure. Once the water level is below the point of equilibrium, the upward force of the pressure begins to win out over gravity, and the net force acting on the water reverses direction, as shown. The pressure acts to slow the water down, finally bringing it to a momentary halt (on the right). But this state of things is just as unbalanced as the one with which we started. The pressure will force the water back up to equilibrium level, but when it arrives there it will overshoot, rising to the original level shown on the left before gravity slows it down and brings it to a halt. From that point the entire cycle repeats itself. If there were no friction, it would continue forever.

In this example, we have all of the elements needed to make a wave. First, some agency causes the system to deviate from equilibrium. Once this deviation starts, the outside agency ceases to be important and the internal forces operating in the system take over. If, as in the case of the water, there are two opposing forces, we find that their interplay produces an oscillatory motion very similar to that of ocean waves.

In point of fact, the kind of mechanism we have just described, in which the forces of pressure and gravity compete and produce an up and down motion of the water surface, is very close to what actually happens in water. The difference between our artificial example and real waves lies in the way the equilibrium of the water is disturbed. The most common source of perturbations in open water is the wind, and the wind blowing across the surface of the water produces a perturbation like that shown on top in figure 6-2. The water surface is ruffled, and the ruffles are blown along in the direction of the wind. When the gravity-pressure competition begins after this sort of perturbation, the straight up and down motion that characterized our simple example no longer holds. Now the water moves in a circular path, as shown on the bottom. The water moves upward on the front edge of the wave, then downward on the trailing edge. The net displacement of the water is still zero—it always comes back to the same point at the corresponding point in each wave—but the actual motion is a little more complicated than in the simplest case. If you pay close attention to the floating piece of driftwood we threw out on top of the waves, you can probably see this slight forward and back motion superimposed on the main up and down path of the wood.

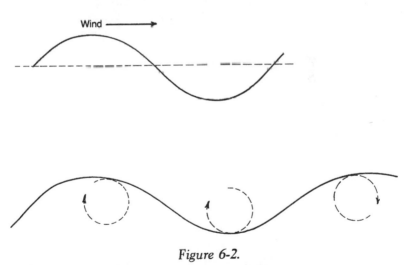

Figure 6-2.

Waves such as those do not stay in one place, but move along in the direction of the wind. Once the wave is started, it will continue to move in the same direction, regardless of whether the wind remains constant or not. The direction and speed of the wave is completely determined by the same interplay between pressure and gravitational forces which causes the up and down motion of the water. The speed at which a wave moves once it is started, in other words, does not depend on the initial disturbance, but only on the properties of the water through which it is moving.

This does not mean that waves formed by the wind will not evolve in response to the wind. The waves which actually result from a given wind depend both on the speed of the wind and the distance over which it blows—a distance known as the "fetch." At first, the wind causes waves of differing lengths to appear. As these build up, they reach a point where the height of the wave is about one-seventh of the distance between crests—a distance known as the wave length. At this height the wave will break, forming a whitecap and giving its energy up to turbulence (we'll talk more about this process in chapter 7). Thus, a wave measuring seven feet between crests will break when it is just a foot high, a wave fourteen feet long when it is two feet high, and so on. The longer the wind blows, the more the shorter wave lengths will disappear and the longer wave lengths will dominate. (The shorter waves disappear because they reach their critical height sooner than the longer ones.) This elimination of the shorter waves will continue until a point is reached where the amount of energy given to the water by the wind is exactly equal to the amount of energy expended in turbulence. When this happens, we say the sea is fully developed. The conditions needed for this to take place depend on the wind speed. For example, a twenty-knot wind must blow for at least ten hours over a fetch of seventy miles to produce a fully developed sea, while a thirty-two-knot wind must blow for twenty-eight hours over a fetch of four hundred miles to reach the same state. Of course, the size of the waves will be different in these two situations—perhaps seven feet in the first and twelve feet in the second. After the wind stops blowing, the waves continue to move in their original direction, giving rise to the smooth motion of the surface known as a swell. The generation of waves on the ocean by winds is sketched in figure 6-3.

This account shows that even for a simple system like wind and water, the way in which waves are produced can be quite complex. But we haven't heard the full story yet; there are other types of waves that can occur in the same system.

Up to now we have talked about only two competing forces, pressure and gravity. But there are other forces at work. For example, we have all noticed that when rain falls on a newly waxed car, it does not spread itself in a smooth sheet over the hood, but forms small beads of water

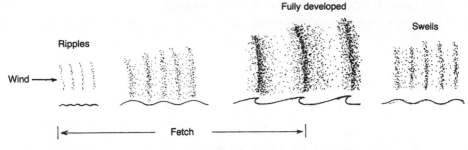

Swells

Ripples

Wind ⟶

|← —————————— Fetch —————————— →|

Figure 6-3.

instead. The reason it does so comes from what is called the surface tension of the water. The easiest way to understand surface tension is to reflect that, while every molecule of water is attracted to every other molecule, those molecules on the surface of the water are subject only to a downward force; all the molecules inside the water are pulling them inward, and there are no molecules outside to counterbalance the inward with an outward force. The surface of any body of water— from an ocean to a raindrop—can be thought of as being somewhat like the surface of a balloon. It will always try to shrink down to the minimum possible size it can have. That is why a raindrop left to itself will form a sphere, which has the minimum surface area for a given volume.

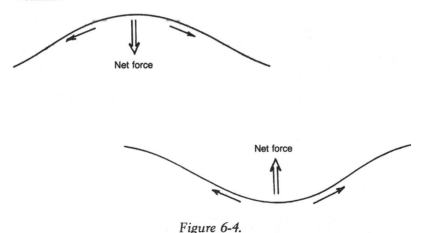

Net force

Net force

Figure 6-4.

It follows that in the making of waves, the existence of surface tension introduces another force which can be used to counteract the effects of pressure. Suppose, for example, that the surface of the water is initially disturbed as shown at the top in figure 6-4. There will be a net force downward on the surface in this case, caused by the forces of surface tension trying to return the surface to a flat plane.(In small ripples, gravitational forces are present but are so small compared to surface tension that they are usually neglected.) By a process with which we are now familiar, this force will push the raised water down, and the energy involved in the push will cause the surface to overshoot, resulting in the configuration shown at the bottom. In this configuration, the force of surface tension will again try to restore the surface to the original flat plane, resulting in a net upward force on the water. It is easy to see how surface tension could produce a wavelike motion in water. Waves created by this mechanism are called capillary waves.

The shapes of real waves are never simple. Note the sunlight glinting on the small capillary waves produced by breezes on the larger waves.

Capillary waves differ from the kinds of waves we have discussed previously in that they tend to have very short lengths—a few inches would be typical. You have probably seen capillary waves many times without realizing it. When the water is relatively calm, a sudden gust of wind will produce a series of ripples on the surface. These ripples are precisely the type of waves caused by the effects of surface tension.

From the nineteenth century until today, the concept of a wave has become more and more central to our understanding of the physical world. Perhaps the best way to make this point is to look at a few other important waves in nature. In each case we will see that the only thing necessary to produce a wave is the presence of a net force that varies in direction so as to reverse the motion of whatever is "waving." In water waves, the interplay between pressure and gravity produced this force, while for capillary waves it was surface tension and pressure.

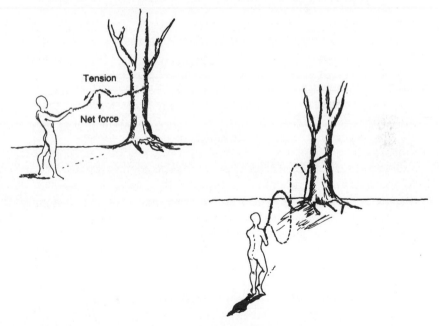

Figure 6-5.

Let us start with an everyday experience. You tie one end of a rope to a tree and hold the other end, keeping the rope taut. A quick flick of the wrist produces a disturbance, which travels down the rope, as shown at the top in figure 6-5. The force at work here is tension in the rope. When the rope is concave downward, the tension also pulls downward, as shown, resulting in a net downward force that accelerates the rope. A similar situation obtains when the rope is below its normal level—tension acts to push it upward. In a rope, then, the force of tension provides the ingredient we need to produce a wave. We should note in passing that a single wave proceeding along a rope as shown in the figure corresponds to a solitary wave traveling along the ocean surface. Such waves are usually caused by an earthquake or volcano eruption. If they are large, as they frequently are, they are known as tsunami, or tidal waves. They sometimes travel over thousands of miles of open sea before striking land—for example, earthquakes in the Aleutian Islands have been known to cause considerable damage in Hawaii and California, over 2,500 miles away.

If you now take the free end of the rope and move it backward and forward at just the right frequency, as shown at the bottom in figure 6-5, you bring about a situation in which various parts of the rope vibrate up and down, but in which there is no lateral movement of the wave. This is known as a standing, or stationary, wave. The sloshing of water in a bathtub is of this type, as are the waves in harbors we talked about earlier. Many large lakes also exhibit this kind of sloshing motion of the water—indeed, the first comprehensive measurements of the phenomenon were made at Lake Geneva in Switzerland. When the standing wave occurs in a large body of water, it is called a seiche.

Standing waves also appear on plucked or bowed strings in musical instruments such as guitars and violins. The vibrations of the string set up corresponding vibrations in the density of surrounding air molecules, vibrations which we perceive as sound. The lower the frequency at which the wave moves on the string, the lower the sound we hear. Since longer strings can have lower frequency waves than shorter ones, instruments that produce low sounds must be larger than those designed to produce high ones. Hence, a bass viol is much bigger than a violin or guitar.

Another tension-induced standing wave effect in musical instruments is exemplified in drums. Here, instead of a string, we have a two-dimensional surface which can be deformed. After being struck, a drumhead can exhibit a two-dimensional standing wave, as shown in figure 6-6. Again, the movement of the drumhead causes the surrounding air to vibrate at the same frequency as the standing wave, a vibration which we hear as the sound of the drumbeat.

Lest you think that standing waves enter your life only through musical instruments, there is a "drumhead" wave that appears on the surface

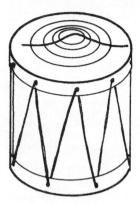

Figure 6-6.

of a fluid in a circular container. Next time you are walking back to your desk or table with your morning coffee, keep an eye on the cup. Most of the time the coffee will just slosh around, but occasionally, especially if you are walking on a hard floor so that the perturbation felt by the coffee is a set of evenly spaced sharp shocks, you will see a series of concentric rings on the surface of the liquid. These rings, sketched on the left in figure 6-7, are actually a standing wave in the coffee. The wave itself is sketched at the right. The surface of the coffee bobs up and down at the crests and troughs of this standing wave, while the points midway between the extremes remain stationary. Light reflecting off the ridges allows you to see the wave. As an aside for those with a technical background, the actual shape of the surface of the coffee is the curve known to mathematicians as a Bessel function, named after Friedrich Bessel (1784–1846), a German astronomer and mathematician.

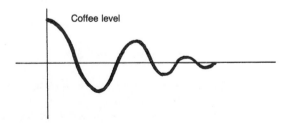

Figure 6-7.

Just as the behavior of ordinary water creates a large variety of waves, so do the effects of ordinary electrical forces. Electrical forces hold atoms together, and electrical forces lock atoms into solids. For example, in a crystal of sodium chloride, ordinary table salt (shown in figure 6-8), every chlorine atom (Cl) has borrowed an extra electron from a sodium atom (Na), so that the chlorine atoms in the crystal have a net negative electrical charge. Similarly, the sodium atoms have a net positive charge, since they've lost a negatively charged electron. A sodium atom, such as the one labeled A in the figure, feels an electrical attraction to each of the neighboring chlorine atoms and is repelled from each of the neighboring sodium atoms. In the equilibrium configuration shown, all of these forces balance, so that there is no net force on (and hence no movement of) atom A. Suppose, however, that some outside agency were to displace this atom slightly to the right, to the position shown by the dotted circle in the figure. It is obvious that in this case the electrical forces will no longer be balanced. The rest of the crystal tries to force the atom back to its original location. Under the influence of the electrical forces exerted by the neighboring atoms, then, atom A moves to the left, picking up speed as it does so. It is moving when it reaches its equilibrium position, so it overshoots and continues to travel toward the left. From the symmetry of the situation, we know that if atom A were to be displaced to the left, the net electrical force would be directed toward the right. Thus, atom A will slow down,

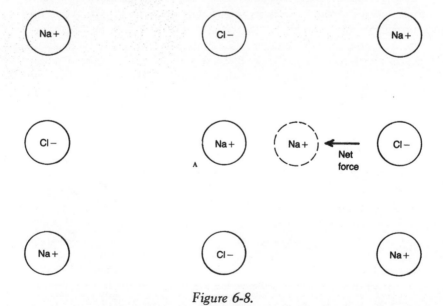

Figure 6-8.

stop, reverse its motion, and proceed to the right again. We have discussed enough different types of waves by this time for this situation to be quite familiar. In a solid like sodium chloride, the electrical force that acts between atoms has precisely the properties needed to produce and sustain waves. These waves aren't as readily available to our senses as are those associated with the surface of water, but we shall see that they are not completely foreign either.

Waves in a salt crystal are just one example of a large class of waves that can propagate in solid matter. In general, the electrical forces allow two kinds of waves: those in which the atoms move back and forth in the direction of the wave, as shown on the left in figure 6-9, and those in which they move in a direction perpendicular to the motion of the wave, as shown on the right. The most familiar appearance of these waves in solids occurs after an earthquake. The quake serves as the perturbation which starts the wave in motion, and once started, the wave moves out in all directions from the earthquake center. It turns out that the two kinds of waves travel through the earth at different speeds, with the compressional wave shown on the left being the faster. Consequently, scientists who monitor earthquakes call this the P (for primary) wave. The wave shown on the right, moving more slowly, arrives at a measuring station after the P wave, and so is given the name of S (for secondary) wave.

Figure 6-9.

The speed of these waves depends on the properties of the material through which each wave is passing. The P wave, for example, will travel about 3.5 miles per second near the surface, but some 10 miles per second deep in the earth. The S wave travels roughly 7/12ths as fast as the P wave. When an earthquake occurs at an unknown location, listening stations around the globe pick up two distinct signals: the time of arrival of the P wave, and the time of arrival of the S wave several minutes to several hours later. By comparing notes between stations, not only can the location and strength of the earthquake be deduced, but also something can be learned about the composition of the material through which the wave has traveled. After all, seismic waves are the only thing we know of that can travel to the center of the earth and come back to tell us about it. Most of the knowledge we have of the earth's structure, in fact, comes from the analysis of various waves detected after the occurrence of earthquakes.

There is a political dimension to S and P waves as well as a scientific one. One of the many technical problems facing people concerned with arms control is the question of whether or not it is possible to distinguish the seismic "signature" of an underground nuclear explosion from that of a small earthquake. This is a very complex issue, but we already know enough about waves in solids to have some understanding of how it might be possible to carry out this task.

Earthquakes are generally caused by the slipping of one side of a fault relative to another. For an earthquake, then, we expect the initial perturbation that starts the wave on its way to be a slipping motion perpendicular to the direction in which the wave will travel. We expect, in other words, that it will be much easier for an earthquake to start an S wave, where the motion of the atoms is in the same direction as the motion of the fault, than it would be for the same disturbance to initiate a P wave. A nuclear explosion, on the other hand, will begin by compressing the rock around it, so we would expect that it would find it easier to start a P wave than an S wave. This difference between the way the two disturbances start should give us a way of distinguishing between them on the basis of the type of waves that arrives at a distant measuring station. To give an oversimplified example, if we saw a P wave arrive without an S wave following, and if the data from several stations allowed us to place the source of the wave in a region where we might expect clandestine nuclear testing to occur, we would have a *prima facie* case for the detection of a treaty violation. On the other hand, if an S wave had arrived and been traced to the same spot, no such case could be made.

The kinds of wave motion we have talked about are but a few of the many examples that physicists have discovered and catalogued over the past two centuries. The root idea is that a wave involves the propagation of a disturbance—for example, a bump on the surface of the water—without involving any net motion of the medium. But no discussion of waves would be complete if we neglected to mention the most ubiquitous waves of all. Known as electromagnetic radiation, these waves include everything from radio waves to X-rays and gamma rays, with

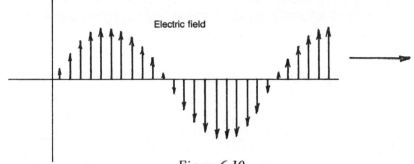

Figure 6-10.

visible light thrown in for good measure.* When it was realized that light had the properties of a wave, similar in many ways to surface waves on water, it was only natural to assume that it was a wave in some sort of medium. Since light seems to be able to travel through the vacuum between stars, it was necessary to imagine some sort of medium permeating all space whose undulations correspond to electromagnetic radiation in the same way that undulations of water correspond to water waves. This medium was called the ether. The idea of such a medium goes back to the Greeks, and it was reasonable to invoke it when seeking a way to explain the behavior of light. The Scottish physicist James Clerk Maxwell, whose work gave us our modern theory of electricity and magnetism, spent a great deal of his short life working out the mechanics of the ether, showing how various kinds of effects could be explained by the stretching, compressing, and twisting of this imagined substance.

But if you think back over our discussion, you will realize that although each wave we described did appear in some medium, the medium was in a sense irrelevant to the description of the wave. In all cases, the wave represented a periodic change in some parameter such as water height or position of atoms. It is the periodicity that is the real stuff of the wave, not the medium through which it moves. Given this fact, it is not too hard to imagine a periodic change that is totally unrelated to an underlying medium. Such a "wave" would be the logical extrapolation of the sort of waves we have been talking about. In the case of light, the quantity that varies periodically is the electric field. A typical light wave is shown in figure 6-10.†

* The evidence for the wave nature of light and other radiation is given in chapter 6 of my book *The Unexpected Vista* (New York: Charles Scribner's Sons, 1983).

† There is actually a magnetic field involved in a light wave, but the essential point can be made by considering electricity alone.

Light, then, represents the concept of the wave in its purest form. It is a periodic change in the electrical field traveling through empty space, without any underlying medium at all. When such a wave arrives at a point in space where there is an electrical charge (for example, at the place where an electron is orbiting in an atom), it will cause that charge to move. The effect of the light wave on the electron can be thought of as being similar to the effect of a water wave on a floating bit of wood—the arrival of the relevant wave causes both the electron and the wood to bob up and down. If the electron happens to be in an atom located in the retina of the eye, then this motion is eventually converted into the sensation of seeing. (Nobody is quite sure how.)

We can close our discussion of waves by returning to the question with which we began the chapter, the question of how it is possible for waves to move toward the shore of a lake in every direction without affecting the level of water in the center. We now understand that this question arises from a basic misunderstanding of the nature of waves. When you see a wave moving across deep water, the essential point to remember is that it is possible for a wave to move in a particular direction without having any net motion of the water in that direction at all.

So the dilemma I encountered on Lake Michigan is resolved, and one more childhood mystery disappears. I suppose I should be satisfied, but sometimes, after a pleasant evening at the beach with friends, I find that if I am just tired enough, and if I've had just enough wine, and if I close my eyes and listen to the surf and think about it in just the right way, I can still see that hole in the middle of the ocean.

7

THE SURF

She struck where white and fleecy waves
Looked soft as carded wool . . .

—HENRY WADSWORTH LONGFELLOW,
The Wreck of the Hesperus

Storms at sea create wind-driven waves. In order to form fully developed waves, the wind may have to blow for hours or even days over hundreds of miles. Once the waves are formed, they will persist in the form of ocean swells over thousands of miles of open sea. When these waves finally encounter a shore, all of this accumulated energy is released in a matter of minutes as the waves change into surf. It's no wonder that the most spectacular display at the beach is the surf breaking against the shore.

The winds that produce the deep water swells can blow in any direction. This means that at any moment swells may be approaching the shore at almost any angle. Despite this fact, anyone who has spent

Waves move toward the beach in a perpendicular direction, regardless of their initial direction. San Gregorio State Beach, California.

time at different beaches is aware that the breakers tend to move in only one direction—parallel to the shore. Why should this be?

After watching the surf for a while, one also notices that waves behave in a variety of ways. Even a single breaker can display different behavior at different points along its front: there may be foam on the crest at one end of the wave at the same time that there is smooth water at the crest on the other end. Why? Like so many of nature's displays, the surf coming in poses many questions for the careful observer.

The Direction of the Surf

Let us begin with the first question. Why does surf usually come straight in to shore, even though the winds that start the waves moving blow in random directions? You might think that the waves generated by storm winds should approach the beach in random directions, so that the direction of their approach today would depend on the direction the wind blew somewhere out at sea a few days ago. That this is not the case results from the character of the deep water wave we discussed in chapter 6. At the surface of a wave in deep water, the particles of water move in a circle. If we watched the particles of water below the surface, we would find that they, too, move in circles, but that the circles get smaller as the depth increases, as illustrated in figure 7-1. Even in waves created by large storms, the motion of the water drops to almost nothing a few hundred feet underwater—which explains why submarines experience smooth "sailing" in storms. A wave moving in water that is more than a few wavelengths deep is called a deep water wave. For all practical purposes we can neglect the effects of the bottom of the sea on the behavior of these waves.

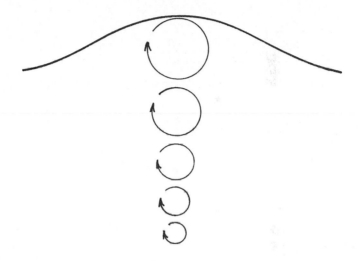

Figure 7-1.

As a deep water wave approaches shore, however, it must eventually enter water that is a fraction of a wavelength deep. At this point the presence of the bottom begins to interfere with the orderly motion of the deeper part of the wave, and the wave's motion undergoes a radical change. Oceanographers say that the wave "feels the bottom." The wave slows down and its shape begins to change. It is converted, in effect, from a deep water wave to the kind of shallow wave we call surf. The important point to realize is that this change has nothing to do with the wave itself; it is the result of the changed environment in which the wave finds itself. A ten-foot wavelength, for example, would be characterized as a deep water wave in a part of the ocean thirty feet in depth, but as a shallow water wave when the depth decreases to five feet.

The mechanism for this radical change in the behavior of the wave

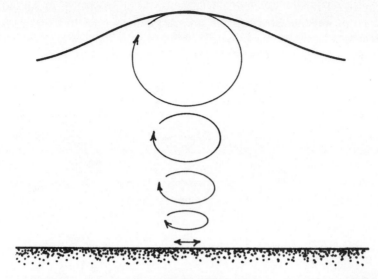

Figure 7-2.

is easy to understand in terms of the underwater motion shown in figure 7-1. If the water is shallow, then the circular motion shown for the deepest part of the wave becomes impossible. The water cannot move up and down, because to do so it would have to dig into the bottom. This means that instead of the free wave shown in figure 7-1, in shallow water the motion of water particles will take on the appearance shown in figure 7-2. The motion of the water at the bottom is necessarily straight back and forth. Just above the bottom, some vertical motion is possible, so the water will move in small ellipses. The farther away from the bottom we get, the less squashed the ellipses become. If the water is roughly half a wavelength deep, the effect will still be very strong at the surface, as shown. It is this distortion of the water paths that slows the wave down. Since the number of waves arriving from the open sea remains the same regardless of what is happening at the bottom, the fact that the waves start to slow down near the shore means that their shapes change as well; the wavelength (the distance between crests) shortens and the water begins to pile up.

To understand why surf always moves in toward the beach, we must keep in mind that the shallower the water, the slower the speed of the wave. Suppose a line of waves is approaching a beach at an angle, as shown in figure 7-3. For the sake of argument, assume that the shoreline is straight and that the bottom drops away uniformly as we move out to sea. When one part of a wave reaches shallow water at point A, it will begin to slow down. The part of the wave at Point B, however, is still in deep water, so it keeps moving along at its usual speed. The result of this situation is that the wave will begin to rotate around the shallow water, as shown. As more of the wave encounters shallow water, this rotation accentuates until the direction in which the crests move has been completely changed. The motion of the wave is very similar to that of a cavalry troop executing a wheel. The soldiers on one side of the line slow down, allowing their fellows to turn around them. The net effect is that regardless of the direction of the waves when they approach the beach, once they feel the bottom they start to turn, and eventually come straight in.

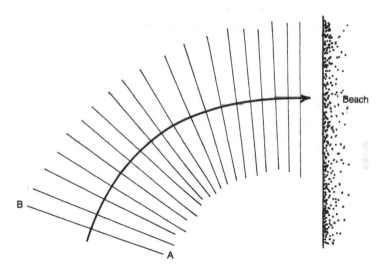

Figure 7-3.

When a wave changes direction upon encountering a region where its velocity changes, we say that it undergoes refraction. Surf is only one example of *wave refraction* in nature: for although you may not be aware of it, you have probably seen refraction countless times in your life. I use the word "see" advisedly, since light, like surf, is a wave and exhibits the same kind of behavior. It, too, changes direction when it encounters an area where its velocity is reduced.

Perhaps I should digress for a moment here to clear up a difficulty many people have when they hear a statement like that. The speed of light plays a central role in the theory of relativity, and the constancy of the speed of light in a vacuum has become one of the essential tenets of science. It is troubling to hear talk of changes in the speed of light. The way to resolve this difficulty is to realize that although the speed of light *in vacuo* must be the same everywhere in the universe, it does not follow that the speed of light in materials is the same as it is in a vacuum. In a transparent material like glass or water, each beam of light progresses in steplike fashion. It travels through the inter-atomic vacuum until it encounters an atom. It is then absorbed by the atom. The atom emits a new beam which, in turn, travels through the vacuum until it encounters another atom, when the entire process is repeated. Thus, the time it takes for light to pass through a pane of glass is greater than the time it would take light to travel the same distance through a vacuum. This has to do with the interaction of light with atoms, not with the speed of light itself. The situation is similar to that of a trans-continental traveler. An airplane can travel at speeds in excess of three hundred miles per hour, but if you have to spend some time in airports changing planes it will take longer than two hours to go six hundred miles.

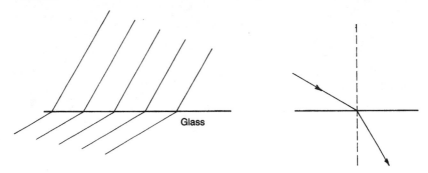

The transition from deep to shallow water waves at a beach is a gradual process, but with light the transition between regions where the wave has different velocities can be quite sharp. At the surface of a pane of glass, for example, the speed of light drops by roughly thirty percent within the thickness of a few hundred atoms. This means that instead of the smooth "wheeling" effect characteristic of surf, the wave front of light entering a medium will suffer a sharp break, as shown on the left in figure 7-4. It is customary to represent this sort of refraction as shown at the right of the figure—instead of drawing the entire wave front, we draw the direction of the wave only, showing the abrupt change in direction caused by the glass surface.

If you have ever used eyeglasses, magnifying glasses, or a camera, you have used the refraction of light, since this is the principle that governs the operation of any lens. The idea is illustrated in figure 7-5. Light from a small object encounters a lens, where its path is bent as shown. This bending is precisely what we should expect from the phenomenon of refraction. The new path of the light rays then brings them to the receiving apparatus—the human eye, for example, or a photographic emulsion. With light coming from the directions shown, the receiver forms the image of the object by extending the light lines beyond the lens, so that it appears to be larger than the object itself. A different arrangement could make the image smaller. The various uses of refraction are the subject of the field known as classical optics.

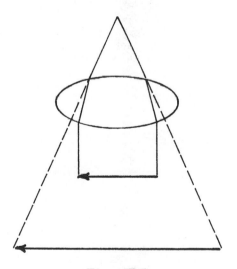

Figure 7-5.

Another interesting refraction effect for light, and one more analogous to surf, is seen in the phenomenon known as the mirage. You have probably seen this effect many times without understanding its cause. It often happens when you are driving on a straight highway on a hot day; the road about half a mile in front of your car takes on a shimmering appearance, almost as if there were a puddle of water on it. When you arrive at the spot, however, the road is dry and the "puddle" appears farther on. The cause of this mirage is the heating of the air by hot road surface. When air is heated, it becomes less dense, and this changes its properties. Light moves more slowly through the warm air near the road surface. Therefore light coming from the sky toward the ground is refracted, as shown in figure 7-6. An observer standing some distance away will see the light coming from the direction shown, and will naturally assume he is seeing something on the surface of the road. The "puddle" is just an image of the sky which refraction has caused to appear away from its normal position. A few years ago, while driving in Montana, I saw an unusual mirage. As the road curved toward a mountain, the "puddle" changed in color from black to dark green. In this instance, the refracted light was not coming from the sky, but from the pine-covered hillside.

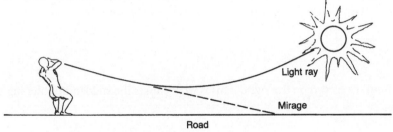

Figure 7-6.

Another aspect of refraction you can easily discover for yourself: it is one that has already started a minor revolution in the communications industry. Next time you are having a candlelight dinner, lift your wineglass and look at the surface of the wine from below. If you lift the glass high enough, the surface appears transparent. As you lower the glass slowly, there comes a point when the surface suddenly shifts from being transparent to being bright and opaque. This, too, is a consequence of the refraction of light waves. Light enters the wineglass from below, as shown on the left in figure 7-7. In the wine, the light's velocity is less than it is in the surrounding air, so when it encounters the wine-air interface it changes direction. Unlilke the surf coming in to shore, however, the wave in this case is moving from a region of low velocity to one where the velocity is higher. Consequently, the wave will be bent in the opposite direction—away from a direction perpendicular to the surface rather than toward it. This path is traced in the figure.

As the glass is lowered, the refracted ray being transmitted through

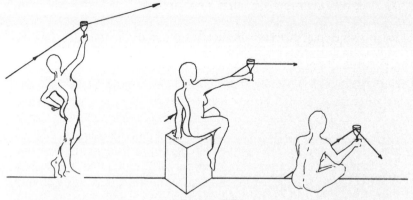

Figure 7-7.

the surface is bent farther and farther away from the perpendicular direction, until finally we reach the critical situation shown in the center of the figure. At this angle, the laws of refraction tell us that the light wave emerging from the wine will move in a direction parallel to the surface. If we lower the glass farther than this, we find the situation shown on the right. The wave encountering the wine-air interface cannot move into the air at all, but instead is reflected back into the fluid, as shown on the right. This is what causes the sudden brightening of the lower surface of the wine as you lower the glass; you are actually seeing the light on the candles reflected off the interface which, at this angle, acts almost exactly like a mirror.

This phenomenon is known as total internal reflection. While it could, in principle, be seen in water waves, it is of practical use only for those waves which, like light, have a relatively short length. Suppose, for example, that a beam of light is directed along a tube of transparent material such as lucite plastic. If the angle between the direction of the beam and the edge of the tube is small enough, then the light beam will undergo total internal reflection when it hits the wall, as shown on the left in figure 7-8. Some simple geometry shows that if a light

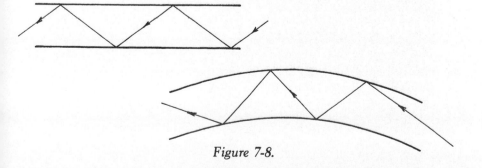

Figure 7-8.

beam suffers total internal reflection in its first encounter with the tube wall, then it will continue to behave in this way in each subsequent encounter as it moves downstream. The net effect of a series of reflections is shown on the right. The light is trapped in the plastic tube and cannot escape, even if the tube is bent. A tube like this is called a light pipe. The only requirement is that the bending of the tube must be gentle enough so that the angle at which the light encounters the wall is below the critical angle required for reflection. Indeed, I have a vivid recollection of a demonstration that one of the pioneers in this field carried out for an audience of physicists. He bent the plastic tube around and showed that the light entering one end always came out at the other. As the demonstration progressed, he twisted the tube into ever more complicated shapes, until he finally wound up with the tube tied in a neat bow, with the light still entering at one end and exiting at the other.

The principle of internal reflection is at the core of the burgeoning new industry of fiber optics. One place where people are likely to encounter it is in the field of medical diagnostics, where it is used to obtain visual images of internal organs. The idea is quite simple. A bundle of plastic light pipes (each smaller in diameter than a human hair) together with a small lamp is brought to the place the physician desires to examine. For example, the tube may be swallowed if it is necessary to view the lining of the stomach. The light which enters each tiny tube is trapped in the tube and carried out to the external world unchanged. There the image is magnified, producing a series of closely spaced dots that the eye integrates into a coherent picture. The formation of a picture by a series of bright and dark dots is familiar to us from television, where a similar process is used. By use of apparatus like this, the physician can examine interior organs, bending light around corners and bringing it out to where he can see it.

A somewhat similar use of light pipes is the subject of experimentation in architecture. One of the major costs of running a large building is providing lighting for the interior. For office buildings, which need intensive lighting only during the day, the use of large windows can help to cut lighting costs. Windows, however, are notoriously inefficient when it comes to insulation—indeed, most of the heat loss in well-insulated structures takes place through the windows. The architect is faced with a dilemma. Increasing window size lowers lighting costs but escalates heating and air-conditioning costs. In areas with mild, sunny climates, these two categories of costs can be comparable, and the design choices are by no means obvious.

Some architects in the southwestern United States have proposed the use of light pipes to resolve this difficulty. The idea is that large mirrors outside the building focus sunlight on one end of a bundle of optical fibers on the roof. The fiber bundles are then snaked down through the building (more or less like electrical cables) to interior rooms, where the fibers fan out over the ceiling to provide uniform lighting. If this idea can be made to work, it will not only provide daytime lighting at lower cost but also a substantial increase in efficiency. After all, an ordinary incandescent bulb converts only five percent of its energy into light—the other ninety-five percent is waste heat, which must be removed by the air-conditioning system. Fluorescent lights, while more efficient, still convert over seventy percent of their input into heat. The light tubes, on the other hand, bring *only* light into the room, so no energy is wasted.

No discussion of fiber optics would be complete without some mention of their use in communication. An ordinary sound wave consists of several hundred wave crests per second traveling through the air. One measure of the amount of information that can be carried by a communication system is the number of wave crests per second that move past a particular point. For example, we might code a message by changing the heights of the waves, so that someone on the receiving end could, with the appropriate instructions, observe the sequence of wave heights and then decode them. To give a trivial example, suppose a wave of maximum height was to be understood as a dot, a wave of half maximum height as a dash. We could send messages by Morse code by this technique.

The problem is that sound waves could carry at most a few hundred dots and dashes per second—perhaps a hundred words. While this is a very high rate compared to normal conversation, it's very low compared to the rate at which a computer can produce and process signals. One way around this problem is to go to waves of much higher frequency. For example, the waves that bring FM radio and TV pictures to your home typically have 100 million crests per second arriving at your antenna. This is why they can carry all of the complex information needed to form a color picture in your TV set. The ultimate extension of this idea, of course, would be to use light where there are 10^{15} waves per second, or 10 million times more than that used to send a TV picture. In principle, then, using light to carry information should be tremendously more efficient than using sound or even microwaves.

Unfortunately, for use on the earth's surface, this theoretical advantace is negated by the simple fact that light is absorbed in the atmosphere while radio waves are not. Light signals simply cannot be broadcast the way that radio waves can be. On the other hand, if we code a light signal and send it somewhere through an optical fiber, there is negligible absorption and the signal can arrive at its intended location undiminished. We can visualize an optical communication system by imagining a "telephone" that converts sound into light waves, which then travel through a network of fibers to the reception point where the signal is reconverted into sound. Alternatively, the initial signal may be electrical impulses generated by a computer rather than sound. Either way, a bundle of optical fibers could carry a large volume of messages—a much larger volume than an ordinary copper wire.

As you might expect, the engineering problems associated with building such a system are prodigious. What may be surprising, however, is that they have largely been solved. Since 1983, New York and Washington, D.C., have been connected by a bundle of glass fibers. The information-carrying capability of these fibers is so high that a single pair in the bundle could transmit the entire *Enclopaedia Britannia* in less than a minute!

The Shape of the Surf

From the simple fact that surf always moves straight in toward the shore we can deduce the laws that govern the new technologies from medicine to communications. But the direction of the waves is in many ways the least interesting property of the surf. Much more complex are the causes of the myriad shapes and variations of the individual waves. A few moments of observation will convince anyone not only that individual waves are different from each other, but that different parts of the same wave may behave differently. In science, complexity in and of itself does not make problems interesting. If the techniques for solving the problem are well known, it just makes them boring. Ocean surf, however, is not one of those problems. The breaking wave is a good example of a system in which simple physical laws operate in such a way as to defy the best efforts of modern science to predict behavior. Even with the best computer models, we cannot at present predict in detail how a wave will break as it moves toward shore.

We can still understand a good deal of what happens to a wave at the seashore, though. As with any attempt to understand a complex phenomenon, the first step is to provide categories into which we can group the events that we see. For incoming waves, there are four such groups:

1. **Plunging breakers**—breakers in which the wave turns over and the top part actually moves forward faster than the wave, eventually forming a tunnel in the water.
2. **Spilling breakers**—breakers in which the foam cascades down the forward slope of the wave.
3. **Collapsing breakers**—small waves moving up a beach so that the front edge of the wave is a perpendicular step.
4. **Surging breakers**—waves that move up the beach without breaking.

These four types of waves are illustrated in figure 7-9.

plunging   Spilling   Collapsing   Surging

Figure 7-9.

The first two are by far the most common. The second two represent rarer phenomena and are usually seen only in sheltered areas or lakes where the surf is less turbulent. The first two types of breakers also share the property that they can be caused either by wind or by the nature of the bottom. Thus, both plunging and spilling waves can be seen on the open sea as well as near shore. In the former, they correspond to whitecaps, in the latter to surf.

The basic process by which a wave breaks is made clear by referring again to figure 7-1, where the motion of individual particles of water is plotted. As a wave approaches the shore, the bottom begins to distort these paths, and eventually the depth become so shallow that the water cannot complete its elliptical circuit. Roughly speaking, this occurs when the depth of the water is 1.3 times the height of the wave. Water which, at greater depths, would rise to fill in the wave is slowed down and even stopped by the presence of the bottom. The result is shown in figure 7-10.

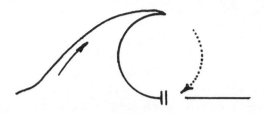

Figure 7-10.

The wave suddenly finds itself unsupported. With the orderly motion of the water disrupted in this way, the wave breaks. Whether the unsupported water at the crest overshoots the wave, producing a plunging breaker, or simply spills down the front of the wave, depends on the details of the wave—particularly the steepness of the wave and the slope of the beach.

A number of analogies have been used to describe the breaking of waves. Some authors refer to the wave as "stubbing its toe" when the water gets shallower. A slightly more accurate image is that of the old vaudeville routine where one comedian is leaning on a cane and the other kicks the cane away, causing the first comedian to fall down. In just that way, the water in the crest finds its weight unsupported and comes crashing down.

By another mode of visualizing we can perhaps make more easily the transition to waves on the open sea. As waves move toward a beach, their shapes change. They start to peak up, becoming steeper as they approach the shore. The reason for this, as we have already noted, is the change in motion caused by the presence of the bottom. We can characterize this peaking process by thinking of two distinct velocities. One is the velocity of the wave—the speed at which the crest approaches the shore. The other is the velocity of the particles of water at the crest of the wave. Normally, the speed of the wave is much greater than the speed of the water. As the wave becomes sharper, however, the speed of the water at the crest increases, eventually reaching a value equal to the speed of the wave. This critical situation occurs when the peak of the wave defines an angle of 120°. It is an unstable situation. If the sharpness of the wave increases beyond this value, the water at the crest will run ahead of the wave and we observe the wave breaking. In deep

Plunging breakers at San Gregorio State Beach, California.

Spilling breakers at Bean Hollow State Park, California.

Surging breakers in an area sheltered by an offshore island. Año Neuvo State Park, California.

water, as we saw, this critical condition arises when the wavelength is seven times the height of the wave. In shallow water, it occurs when the depth is 1.3 times the height. A thirty-foot wave will break when the depth of water is about thirty-nine feet, while a three-foot wave will break when the bottom is slightly less than four feet below. The behavior of the surf will thus depend on the size of the incoming waves, which, as we saw in chapter 6, depends on the wind that formed them. Nor is it unusual for conditions to be such as to cause a wave to break more than once. For example, a thirty-foot wave may break far from shore, creating in its aftermath smaller waves, which continue moving inward and which may themselves break, creating still smaller waves that repeat the process.

The foam marks the spot where waves break. In this photo, the waves reform several times before coming to the beach. San Gregorio State Beach, California.

Plunging breakers need not be large. Here, with two bull elephant seals fighting in the background, we see them less than a foot high. Año Neuvo State Park, California.

Geysers formed by compressed air created when the wave moves against the rock. Bean Hollow State Beach, California.

Two plunging breakers arrive almost simultaneously in this cove at San Gregorio State Beach, California.

If you stand on the shore and see waves breaking out to sea, you can be sure that the bottom at that point is shallow enough to meet the requirement for a breaking wave. If, as often happens, you see offshore breaking at only one point along a beach, you can conclude that there is an underwater hilltop or ridge at that point. This process of breaking and reforming of waves in response to the changing bottom can be repeated many times as the wave moves in; it is not at all uncommon for the last plunging breaker to be less than a foot high when it finally reaches the beach.

This dicussion of the general properties of the surf takes care of the most conspicuous features but ignores what, to me, is the most fascinating aspect of the subject. While general categories of breaking waves can be defined, a few moments' observation wil show that no two waves coming into a beach are exactly the same. All sorts of variations occur, each obeying the general laws we have outlined, but each arising from slightly different conditions in the history of the wave. For example, a plunging breaker can come in many different forms, depending on the amount by which the velocity of the water exceeds the velocity of the wave. Some examples are shown in figure 7-11. The overturning water may strike in front of the wave or on the forward face. When it hits, it may simply produce a lot of foam or it may splash. Air trapped in the hollow of the wave may be compressed, giving rise to geysers of water when it escapes—usually when the wave hits a rocky shore. If the shoreline has slight irregularities, waves can arrive at the beach from different directions, creating interesting effects such as one set of breakers overtaking another. A wave coming in to shore at an angle may even show spilling behavior in one area and plunging behavior in another. It is this infinite variety, no doubt, which makes the beach such a fascinating place.

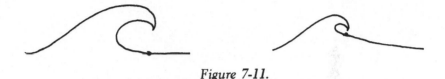

Figure 7-11.

I like to think of wave-watching as being akin to listening to music. To the beginner, each performance of a Chopin sonata sounds the same as the next, but as one attends more closely, the differences of detail begin to appear. After a while, you are aware not only that every artist produces a different version, but also that different performances by the same artist are never identical. So it is with waves. They all look like moving walls of water—at first. With time, the subtle differences begin to be noticeable and we recognize each separate wave as an individual work of art.

8

BUBBLES IN
THE FOAM

You know, of course, if you're not a dunce
Why it went to pieces all at once
All at once and nothing first
Just as bubbles do when they burst
End of the wonderful one-hoss shay
Logic is logic, that's all I say

—OLIVER WENDELL HOLMES,
"One-Hoss Shay"

The final result of the generation of waves at sea and the destruction of those waves in the surf is the foam left on the beach. Spray and foam are interesting, if only because they are so ubiquitous. Oceanographers estimate that at any moment three to four percent of the ocean surface is covered with foam generated by breaking waves, either in the open sea or on shorelines. This means that an area of the earth's surface roughly equivalent to the North American continent is composed of the same tiny bubbles that wash up along your favorite beach. It would be amazing, given the importance of the ocean in generating the world's climate, if foam didn't play some role in our daily lives.

Besides, bubbles and droplets are interesting in and of themselves.

At any moment, 3 to 4 percent of the earth's surface is covered with foam like this.

They are objects we can see with the naked eye, but which reflect the forces that act at the atomic level. The liquid skin of a bubble may be no more than a few hundred atoms thick, so that the details of the inter-atomic force may be seen. These details are often obscured in bulk matter, where the number of atoms involved in producing a visible effect often approaches 10^{26}.

Actually, the existence of bubbles leads to a rather more fundamental question: Why do lumps of liquid stick together at all? Why, given that water is made up of atoms that are electrically neutral, do forces exist that make it form into drops? If we bring this question down to the most fundamental level, how can two neutral atoms exert a force on each other when the laws which govern the electrical interaction say that only charged particles can exert an electrical force?

This is an interesting conundrum. In fact, it is a favorite question of mine—I often use it when called upon to sit on the committee that hears a graduate student defend his or her thesis before getting the Ph.D. The practice isn't as sadistic as it sounds. The Ph.D. exam is no longer the do-or-die institution it once was; the student's adviser has invariably seen that the work is of the quality needed for the degree, and as a result virtually no one ever fails the exam. I usually tell my students that the purpose of the defense is to make sure they don't think they know everything when they leave the university. For this purpose the question is admirably suited.

Think first of two atoms, which may or may not be identical. Each is, as we know, electrically neutral. But we also know that this electrical neutrality arises because the positive electrical charge residing at the center of each atom is exactly canceled by the negative charge which resides in the electrons in their orbits. When the atoms are far away from each other, this overall electrical neutrality means that there is, indeed, no force between them. The force exerted by the electrons is canceled by the force exerted by the nucleus. When the separation between the two atoms is less, however, this need no longer be true. The force exerted by a charge depends on the distance from the charge to the point where the force acts, and when atoms are close together, the fact is that the distance from a point in one atom to the electrons in another may be very different from the distance from that point to the nucleus. The net force, therefore, need not be zero.

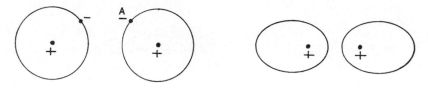

Figure 8-1.

An example is shown in figure 8-1. An electron at the point labeled A in the right-hand atom will feel a repulsive force due to the electrons in the left-hand atom. This force will tend to move the electron at A to the right. The electron will, of course, feel an attractive force due to the nucleus of the left-hand atom as well, but the fact that the nucleus is farther away from point A than are the electrons will mean that the repulsive force wins out. A similar argument, of course, can be made for the electrons in the left-hand atom. The net effect of bringing the two atoms close together, then, is shown on the right of the figure. The electron clouds are distorted from the symmetrical shape they take when the atom is isolated. The center of the cloud shifts away from the nucleus in response to the presence of the neighboring atom. In the jargon of physics, we say that the atom acquires an induced dipole moment.

Thus, although atoms do indeed have an overall electrical neutrality, the fact that the distribution of electrical charge inside of the atom is not uniform means that distortions such as the one we've just discussed can create a situation in which forces might exist between atoms at short distances from each other. In fact, if we follow this idea mathematically, we find that the force between two atoms due to distortions such as those shown in the figure is attractive—that is, it tends to pull the atoms toward each other—and vanishes rapidly as the separation between the atoms is increased. If the two atoms are identical, we term the force generated in this way the "force of cohesion." This is the force that holds drops of liquid together. If the two atoms belong to different chemical species, we speak of the "force of adhesion." This is the force that makes one substance stick to another.

Having seen how electrically neutral atoms can attract each other, it's not hard to extend the idea to more complex electrically neutral objects like molecules. A molecule of water, for example, consists of one atom of oxygen and two of hydrogen, the familiar H_2O. However, arguments similar in character to (but slightly more complicated than)

Figure 8-2.

the ones given for single atoms lead to the same conclusion: there is a force which tends to hold water molecules together when they are closely packed, and a force which tends to make molecules of water adhere to other substances. It is the latter force which makes the water droplets stay on your skin after you climb out of a pool or a bathtub.

The force of cohesion is also responsible for the presence of bubbles and droplets in the surf-generated foam at the beach. Let's start with the bubbles. Imagine, if you will, an idealized membrane just one atom thick, as shown in figure 8-2. Each atom will be attracted to its neighbors, so that the forces on it will be along the membrane, as shown. Obviously, since the membrane does not move, the force to the left on any given atom must be equal to the force to the right. The force per unit length acting along the surface in this way is called the surface tension of the membrane (see chapter 6).

It is clear that the existence of surface tension depends on the atomic forces of cohesion. It is also clear that surface tension will act to pull the membrane inward, causing it to shrink down on itself. The shrinking will be stopped only when the pressure of the material inside the bubble (usually air) is large enough to balance the forces of surface tension and produce a stable equilibrium.

The situation in a droplet is quite similar. A molecule inside the drop will be pulled equally in all directions by its neighbors, and hence will feel no net cohesive force. A molecule in the surface, however, will feel two forces. Both forces arise from the same inter-atomic reaction we have been discussing, but because of the arrangement of the atoms in the drop they act in different directions. One force, exactly like the one in the membrane, is directed along the surface. The other results from the fact that molecules in the liquid are exerting an inward force on the surface molecule—a force which cannot be balanced by molecules on the other side of the surface because, by definition, there are no molecules there. Both of these forces tend to pull the droplet in on itself until they are balanced by the pressure of the liquid in the drop.

Thus, to all intents and purposes, we can regard the forces at work in a bubble and a droplet as being identical, even though the two systems appear to be rather different. The reason for this somewhat surprising state of affairs lies in the fact that the essential characteristics of both structures have to do with the properties of surfaces. What the surface encloses is a good deal less important than the existence of the surface itself. It will, in fact, be useful to think of both bubbles and droplets as being analogous to balloons, another system that is "all surface." It is necessary to exert a force to expand the balloon because we have to increase the distance between surface molecules, doing work to oppose the attractive force between them. Left to itself, a balloon will assume a spherical shape and collapse until the internal pressure balances the forces of surface tension, behaving just like the droplet and the bubble.

Foam

The frothy residue left by incoming waves is a mixture of air and water known as a foam. The air enters the foam through the turbulent action of the surf, which mixes water and air together. The ocean foam, then, is created in much the same way as a cook makes whipped cream, by forcing air into a liquid through mechanical action. Chemists classify other kinds of foams according to the way that the gas is introduced into the liquid. Sometimes this mixing is done chemically. One "foam" created in this way is common in many kitchens—the action of yeast in bread dough creates bubbles of carbon dioxide that cause the dough to rise. The mixing can also be done by allowing gas to be absorbed in a fluid under high pressure and then lowering the pressure so that the gas forms bubbles. This is the technique used to produce the "fizz" in carbonated soft drinks and beer. So, beer, bread, whipped cream, and a variety of other forms are all similar to what you see at the beach.

Up to this point, we have given a rather idealized picture of the structure of bubbles in the foam. As you might suspect, the real world is more complicated than this picture suggests. For example, although it is possible to mix air and seawater to produce a foam, pure water (or any pure liquid) will not foam at all. It is necessary that a foaming agent be present in the liquid. This is a fact of which we are all aware—it's very difficult to make a foam by rubbing your hands in tap water, but the addition of a little soap or shampoo changes the situation radically.

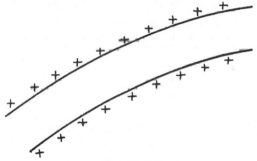

Figure 8-3.

The role of the foaming agent is to introduce extra electrical charges into the bubble membranes. A cross section of a membrane is shown in figure 8-3. The extra electrical charges serve to provide a repulsive force which prevents the membrane from getting too thin and which therefore increases the lifetime of the bubbles.

The Rainwater Recipe

In chapter 2 we saw that the incorporation of ocean minerals in rainwater formed an important link in the chemical reactions that make the sea salty. We also saw that the chemical composition of the river water that flows into the sea was markedly different from the composition of the sea itself. It should come as no surprise, then, to learn that rainwater differs in composition from both the rivers and the sea.

It was John Dalton, the man who gave us our modern theory of the atomic structure of matter, who first showed that there is salt in rainwater. That was in 1822. For a century and a half after this discovery, the question of why rainwater has the composition it does remained a mystery. We have already seen how rainwater leaches ground minerals to provide the chemicals that flow into the sea in rivers, and how chemical reactions in the sea remove almost everything except sodium and chlorine in the form of salt. The question of the composition of rainwater, then, is the only missing link in our understanding of the water cycle on the earth.

In 1963, New Zealand scientist G. A. Dean provided hard data for the analysis of this question by undertaking a meticulous analysis of rainwater falling on a small town on the coast of his country. The result of his work was a "recipe" for making rainwater, which follows:

Rainwater à la Dean

.5 milligrams ordinary seawater
4 milligrams dried plankton and algae
enough distilled water to make one liter
Mix all ingredients together and stir well.

The first ingredient does not surprise us—we already know that some normal seawater must exist in the rain. It serves the purpose of supplying the correct ratios of sodium, chlorine, and calcium, as given in the table on page 13. The second ingredient contains the surprise, because the dried organic matter that Dean found represents a concentration roughly two thousand times that found in the ocean or in evaporated seawater. Somehow the way that water from the sea is incorporated into rain must include a process that enhances the amount of organic material.

A number of such mechanisms have been proposed, but the most interesting from our point of view is one put forward by oceanographer Ferren MacIntyre of the University of Rhode Island. He pointed out that the algae and plankton that, according to the rain recipe wind up in rainwater, tend to congregate at the ocean surface where they can make use of the incoming energy in the form of sunlight. Thus, we would expect the concentration of organic matter in a thin layer near the ocean surface to be much higher than it would be in seawater in general. This organic material will not, of course, be taken into the air by evaporation, but there is another phenomenon which can do that particular job: the bursting bubble.

We have already mentioned the large amount of the earth's surface that is covered with foam at any given moment. If you watch the foam at a beach, you can quickly convince yourself that most of the bubbles in it will burst in less than a minute—what, after all, is more transient than a bubble? As it happens, the way in which bubbles break has been a favorite field of study for people developing high-speed photographic techniques, so we know a good deal about it. The statement by Oliver Wendell Holmes quoted at the start of this chapter not withstanding, bubbles do not break "all at once," but follow a well-defined sequence,

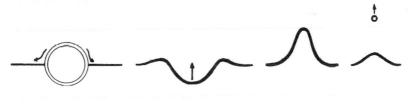

Figure 8-4.

which is sketched in figure 8-4. When a bubble first rises above the water level, gravity starts to pull the fluid out of the bubble wall, as shown on the left. When the wall is sufficiently weakened, it will break at the top, at which point the submerged surface, under the influence of the force of surface tension, will begin to snap upward, as shown in the left center. The surface will overshoot when this happens, forming a thin column (right center), which then breaks into one or more drops (right). If you watch closely, you can see an exactly similar sequence of effects on the water surface after you drop a pebble into a pond. The bubbles formed this way can be thrown a foot or more into the air, allowing plenty of time for the water in the drop to evaporate, leaving behind an airborne particle which will eventually be returned to earth in rain.

MacIntyre estimates that a single bubble with a diameter equal to the thickness of a dime can, when it bursts, inject 30 billionths of a gram of salt and .3 billionths of a gram of organic matter into the atmosphere. This may seem like a small amount of material, but if we add up all of the bubbles in the ocean, we find that several billion tons of salt per year can be injected into the atmosphere by this mechanism, together with several hundred million tons of dried organic material.

If you spend time near a beach, you have direct experience of the presence of organic material in the air. A day's walk along the beach will often leave you feeling that your skin and hair are greasy. In fact, they are—you have been walking through a very thin soup of airborne plankton and algae. Similarly, your car windshield may be greasy if it has been parked facing into the wind. Some of the most luscious artichoke fields I have ever seen are located within a few yards of the Pacific south of San Francisco—fields where the plants obtain a good deal of their nourishment from the same airborne material. I have been told that meat from sheep that graze near the coast in Normandy has a distinctive flavor because of the salty grass found there, and that this flavor is highly prized by the French. If you think of multiplying these local observations by the vast amount of the ocean's surface covered with foam, it is clear that bubbles play a larger role in the world ecology than their size and evanescence suggest.

Bubbles in the Beer

One of the best ways to observe the behavior of bubbles is also the most accessible. A glass of any carbonated beverage—soft drink, beer, or champagne—can, with a little knowledge, be turned into a marvelous laboratory. These foams are created differently from that of the ocean, of course, but we now know that once a bubble is formed its behavior is dominated by surface tension. Whether it was formed by the trapping of air in a plunging breaker or by the sudden lowering of pressure when a bottle is opened is irrelevant. A bubble is a bubble.

A carbonated drink can be made in two ways. Molecules of carbon dioxide (CO_2) from an external source can be forced into a liquid at high pressure. Alternatively, the chemical reactions initiated by yeast can make the carbon dioxide inside the fluid without any outside intervention. In both cases, the carbon dioxide is kept in the fluid by high pressure. Soft drinks are made the first way, champagne and some beers the second. The high pressure is maintained, of course, by the can or bottle in which the fluid is stored

When the bottle or can is opened, the pressure drops as the imprisoned gas is released to the atmosphere. This outrush of gas is the pop you hear when the cork comes out. The release of pressure allows bubbles of carbon dioxide to form. It is these bubbles that make the familiar fizz in the drinks.

If you hold a glass of champagne up to the light and look at the surface, you can actually see the process that moves organic material from the ocean to the air. As bubbles come to the surface and burst, you can see tiny bits of spray being thrown up from the surface. If the light is just right and the bubbles are big enough, you can also see evidence for the draining of liquid from the bubble membrane prior to breaking. Just before the bubble bursts, try to detect an iridescence in the bubble surface. You should be able to see a kind of violet or blue-green luster where the light hits the bubble. This iridescence is caused by the light encountering a thin film of material. Some of the light is reflected at the upper surface, some is transmitted through the film and reflected at the lower surface. The recombination of the two waves is what produces the visible effect, but in order for this to be seen at all the film must be thin enough so that the second wave isn't absorbed during its transit through the liquid. The onset of iridescence marks the point where the thinning out of the bubble wall begins as it can no longer resist the little shocks and irregularities to which it is normally subjected. Sooner or later, a shock will come along which the bubble wall is no longer strong enough to resist and the process of bursting we described above will begin.

A carbonated beverage is known in technical terms as a supersaturated system. The amount of carbon dioxide contained in the fluid amounted to the total that could be absorbed when the pressure was high. As soon as the bottle is opened, the pressure suddenly drops, and the fluid is in a situation where it actually contains more carbon dioxide than it can hold at the new, lower pressure. The mere fact that this is so, however, does not mean that the gas will immediately come out in the form of bubbles. A perfectly unperturbed system can stay in a supersaturated state for quite some time—in principle, it could remain supersaturated forever. On the other hand, any real system will have irregularities which serve as nuclei for the formation of bubbles. The act of opening a bottle and the subsequent outrush of gas creates enough turbulence in the fluid to set off the process of bubble formation. This usually happens too fast for the eye to follow.

There is another phenomenon common to carbonated beverages where the process of bubble creation through nucleation can be observed. After the drink has been poured into a glass, you often see a steady stream of bubbles rising from one point. Typically, the bubbles are a half to a quarter inch apart. What is happening to produce these bubbles is shown in figure 8-5. The spot which serves as the initiation point for the bubbles is actually a tiny crevice in the otherwise smooth glass surface. When the drink is poured, a small bubble of air is trapped in this crevice, as shown on the left. A molecule of carbon dioxide in the fluid near the air is subject to two forces: an attractive force tending to keep it in the fluid, and an attractive force tending to pull it toward the air pocket. The first force arises through the interaction of the molecule with water molecules in the fluid; the second through the interaction of the molecules with the oxygen and nitrogen in the pocket. It turns out that, because of the structures of the atoms and molecules involved, the second force is stronger than the first. Consequently, molecules of carbon dioxide will leave the liquid and stick to the outer surface of the air pocket, as shown in the center of the figure.

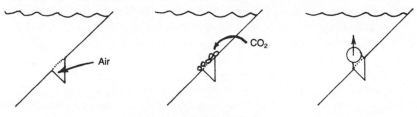

Figure 8-5.

As more molecules are pulled out of the liquid, a bubble begins to form. this process is aided by a new force—the attraction between molecules of carbon dioxide in the fluid and molecules of the same gas in the surface film. Thus, the more material a bubble has, the more it will attract. Eventually, we have a situation like that shown on the right. A large bubble of carbon dioxide has formed and a new force—the ordinary force of buoyancy—comes into play. It tends to push the bubble upward, and when the bubble is large enough, buoyancy overcomes the force of adhesion holding the bubble to the air pocket and the bubble starts to rise to the surface. The air pocket, now "clean" once more, is free to start the whole process again, giving rise to the steady stream of bubbles we observe.

There is something about watching bubbles that inclines the human mind toward philosophical musing—a point emphasized repeatedly by writers of popular songs. Yet contrary to what you might expect, there is at least one historical incident in which such musings led to a very important scientific result. In 1952, Donald Glaser, a physicist at the University of Michigan, was engaged in just this activity at a campus watering hole when he realized that what he was watching could provide the key to a new way of detecting the behavior of elementary particles. These particles, which are smaller than the nucleus of an atom, cannot be seen with the naked eye, but because they are thought to contain the key to the ultimate structure of matter, studying their properties is an extremely important scientific activity. What Glaser realized that evening in Ann Arbor was that *anything* could be made to serve as the nucleus for bubble formation in a supersaturated liquid, even the debris left behind when elementary particles collide with ordinary atoms. This became the working principle of a device called the bubble chamber, and its invention won Glaser the Nobel Prize in 1960.

The idea is this: When an elementary particle passes near an atom, the electrical interactions may be strong enough to tear an electron out of its normal orbit. The resulting atom, known as an ion, has one too few electrons and therefore possesses a positive electrical charge. If the atom happens to be in a supersaturated fluid, then molecules of the gas will be attracted to it, much as they are attracted to the air pocket in a champagne glass. A bubble will grow—a visible bit of macroscopic evidence that an elementary particle was present.

In a working bubble chamber, liquid is kept at high pressure in a chamber which has a movable piston on one side. After a burst of particles from an accelerator has entered the chamber, the piston is pulled down, lowering the pressure and allowing bubbles to form. Next, a set of cameras with flash attachments photographs the bubbles, recording whatever happens in the chamber. The bubbles are then swept from view by magnetic fields, the pressure restored by raising the piston, and the entire process repeated. In modern high-energy physics experiments, chambers several feet on a side filled with liquid hydrogen at a few degrees above absolute zero are cycled every few seconds, producing hundreds of thousands of photographs to be analyzed. They have been a major tool in our investigation of the properties of the subnuclear world.

There is another field of science that is touched by the behavior of bubbles. Although we have talked about bubbles solely in terms of fluids, it is clear that the only things necessary to produce bubble-like behavior are two competing forces. One of these, analogous to surface tension, must act along a surface to pull the system together; the other, analogous to pressure, must act to push it apart. If two such forces exist in a system, then it ought to exhibit behavior similar to bubbles, even though the system itself may seem the farthest thing from a fluid mixture of gas and fluid.

One of the facts that radiochemists and nuclear physicists have discovered about the world is that although there are almost a hundred naturally occurring elements, there seems to be a limit to the mass of the nucleus of atoms. Uranium, with 238 particles in its nucleus (92 of them positively charged protons), is a fairly common element in nature. It is, in fact, more abundant than familiar substances such as silver and mercury. But heavier elements than Uranium are either rare or nonexistent in nature, and are known only because small amounts have been produced artificially in laboratories or nuclear reactors. The reason for this fact is actually quite straightforward: nuclei of atoms heavier than uranium are unstable, decaying relatively quickly into fragments which are the nuclei of lighter atoms. Even uranium is unstable, but with a half-life of 4.5 billion years, it can be thought of as representing the borderline between stability and instability in the nuclear world. To understand the absence of the trans-uranium elements in nature, then, we have to understand the instability of heavy nuclei.

The most surprising thing about the properties of bubbles and droplets is that the question of nuclear stability—seemingly as far as we could get from the product of breakers on the shore—was first solved by the use of a model in which the nucleus itself was treated as a tiny droplet. Going by the name of the "liquid drop model" of the nucleus, this approach pictures the behavior of the nucleus as being determined by the action of two competing forces. One of these, the analogue of pressure in a drop or bubble, is the electrical repulsion between protons in the nucleus. Since all protons have a positive electrical charge, there will be a force between any two of them tending to push them apart. The second force, the analogue of surface tension, is supplied by what physicists call the strong interaction. Discovered in this century, this is the force which provides the "glue" that overcomes the electrical repulsion and holds a nucleus together. For our purposes we can imagine that it is just like the attractive force that exists between atoms. The strong force affects neutrons and protons alike, and leads to surface tension in the nuclear "drop" in exactly the same way that the cohesive force between atoms operated in figure 8-2, page 101.

When we think of the nucleus in this way, we realize that it represents the smallest bubble in nature. We can think about its stability in just the same way as we can think about the stability of bubbles in the surf or in our beer. We know that if a bubble is unstable, it will break, and we know that one way to break a bubble is to increase the pressure of the gas inside it. By analogy, then, we might expect that increasing the pressure in the nuclear bubble by adding protons will cause that bubble to break as well. But when we add protons to a nucleus, electrons will attach themselves to the atom to balance the additional positive charge, and the end result is a new chemical element of higher mass than the one with which we started. Just as pumping air into a bubble will eventually render it unstable, adding protons to a nucleus will eventually result in an unstable system as well. Simple calculations indicate that the instability sets in when ninety to ninety-five protons are present— just the limit we observe in nature. Thus, a rather fundamental problem in nuclear chemistry can be resolved by thinking about the behavior of bubbles.

I have cause to appreciate this fact. The first time I taught the graduate course in fluid mechanics at the University of Virginia, a student by the name of Bruce Broulik (now a professional physicist) and I decided to use the liquid drop model of the nucleus to investigate an interesting question. The strength of the strong interaction is one of the fundamental constants of nature, and it's important to know whether it is really constant or changing with time. Bruce and I reasoned that if the strong interaction had been stronger in the past than it is today, then the surface tension of nuclei would have been greater then than it is now. This would imply that some nuclei which are just over the stability borderline today would have been stable in the past, and this, in turn, would imply that some nuclei not found in nature today would actually have been present. The absence of these borderline nuclei, therefore, would be evidence for the constancy of the strong interaction in time.

We discovered that a particular isotope of plutonium, ^{244}Pu, was not found in nature, and we used this fact as the keystone of a brilliantly written and cogently argued paper showing that the strong force was, in fact, constant over geological times. We were still congratulating ourselves on our cleverness six months later when a team of geochemists announced the discovery of traces of ^{244}Pu in a block of ice taken from Antarctica.

I tell this story partly because it illustrates the far-ranging physics inherent in the simple bubble, but also because it illustrates an important difference between science and other technical subjects like mathematics and logic. In science it is possible to start from a set of reasonable hypotheses, follow correct reasoning to impeccable conclusions, and still turn out to be wrong. This is because in science there is always an impartial external criterion against which to judge results, the criterion of experimental validity. Scientific theories, in other words, not only have to be self-consistent, they have to be verified by concrete evidence.

9

THE SHAPE
OF THE WAVE

*(For I can) change shapes with Proteus to advantage and put
the murderous Machievel to school.*

—WILLIAM SHAKESPEARE,
Henry VI, PART III

The trouble with ocean waves is that once you get past the regularities
of the surf and enter deep water, they stop looking like the textbook
examples we've been talking about. Deep water waves are seldom simple
undulations of the ocean surface. More often the waves appear to be
almost triangular in cross section with large and small irregularities on
their surfaces. Furthermore, if you watch them closely, you will note
that they change shape as they move along. How such irregular, con-
stantly shifting forms arise out of simple regular waves is one of the
most fascinating stories in physics.

In chapter 6 we saw how ocean waves are generated by winds blowing
for long distances over the surface. The net result of the wind will be

ocean swells—smooth, regular waves traveling away from the windy area. If only one wind blew on the ocean at a time, we would not expect to see a great deal of irregularity in wave shape. But, as we show in figure 9-1, the more normal situation involves two or more wave fronts, each generated by a different wind, arriving at the point where we observe the surface. To understand the outcome of this sort of encounter, we have to know what happens when two waves come together at a single point.

The best way to visualize the situation is to imagine that as a wave comes to a certain point, it gives a command to the surface—"Move up three feet," or, "Come down ten inches." By thinking in this way, the question of what happens when two waves arrive simultaneously at a point is solved: The surface "adds up" the commands it receives from the separate waves. In our example, the surface will take 3 feet, subtract 10 inches, and rise to a height of 2 feet 2 inches. This particular process is called interference. We can get some idea of how it works by looking at other examples.

In the case of waves of equal height and wavelength, the height of the surface will depend on which sections of the two waves happen to come together at the point of observation. If two crests arrive together, then the height of the water will be twice that of either wave taken individually. If two troughs arrive together, the water level will be

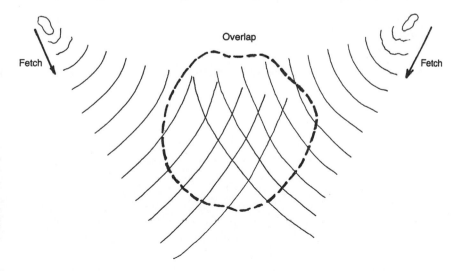

Figure 9-1.

depressed twice as much as for a single wave. If, on the other hand, the crest of one wave arrives at the same time as the trough from another, the water level will not be displaced at all. The surface will receive two "commands" which cancel each other out.

There are two ways to see the results of the interference in this case. One is to look at the surface of the water in the area of overlap at a fixed time—in effect, to freeze the surface with a photograph and study it. In this case we will see a situation like that shown in figure 9-2. At points like the ones labeled A and B, where wave crests have come together, the water level will be twice as high as normal. At other spots, like those labeled C and D, where a crest of one wave meets the trough of another, the water will be unperturbed. There will also be other spots where the water level will be twice as low as normal. These three cases represent extremes, of course. At most spots the surface will be at some intermediate level, corresponding, for example, to the crest of one wave coinciding with a point halfway up the crest of another.

Two factors should be stessed about this illustration. One is that the water surface would not look the same in photographs taken at two different times. As the two waves move through the region, the positions at which crests and troughs coincide will change. Thus, someone watch-

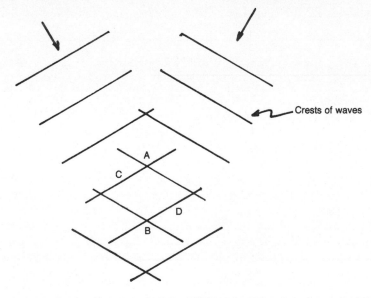

Figure 9-2.

ing the area will see the water rise and fall regularly at each point. The second factor has to do with the shape of the waves in the overlap region. In figure 9-2, we have shown a situation where both waves entering the region are smooth, fully developed swells. Such waves have essentially the same cross section at any point along their front. Once they encounter another wave, this is no longer the case. For example, in figure 9-2, point A is on the crest of a wave advancing from the left. Point C is at the point where the crest of the same wave would be, but, because of interference, the water at C is at its undisturbed level. Thus, even in this simple example we see that interference between waves can entirely change the shape of the wave.

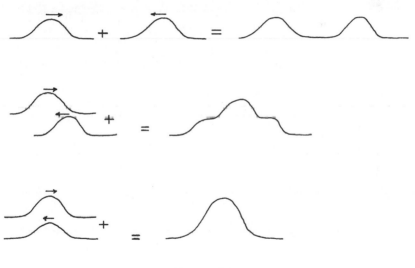

Figure 9-3.

Another way to see that this is so is to consider what happens when two waves collide—to substitute a moving picture of a single wave for a photograph of many waves. The situation would then be like that shown in figure 9-3. As two waves approach each other, the surface of the water takes on a series of complex shapes. Only when the two waves are halfway through each other—when the crests and troughs are aligned as shown—will the water surface have a simple shape. We see that even in this simple case the water surface can take on rather convoluted shapes because of interference.

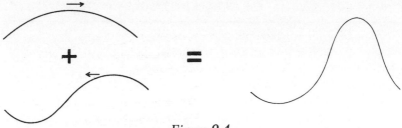

Figure 9-4.

The principle of wave interference can easily be applied to more complicated situations. For example, we show in figure 9-4 two waves of unequal heights and wavelengths passing through each other.

Again, quite complex surfaces can be produced even by two simple waves.

There is still another way of representing this surface, shown in figure 9-5, that will be very useful later. We know that despite the complexity of the shape, the surface is actually nothing more than the sum of two waves. The wave coming from the right has twice the height and half the length of the one coming from the left. If the two waves have the same velocity, then the difference in wavelengths can be translated into the statement that the right-hand wave has twice the frequency of the

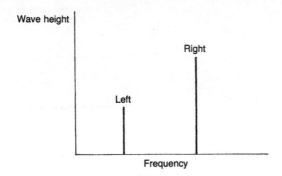

Figure 9-5.

one on the left, since two of its crests will move past a point in the time it takes one crest of the left-hand wave to do so. The alternate way of representing the surface, then, is with a plot like that shown in figure 9-5. There each of the waves is represented by a vertical line. The height of the line corresponds to the height of the wave, and different lines correspond to different frequencies.

When I say that this graph is an alternate way of representing the surface, I mean that given this graph, we could reproduce the two waves shown in figure 9-4. From these waves, we could use the rules of wave interference to reconstruct the surface. The equivalence of figures 9-4 and 9-5 will be very important in our later discussions.

The Real Ocean

An observer in a ship in the middle of the ocean will observe the intersection of many more than two waves. Swells generated by storms and winds in many parts of the ocean will be coming together, each swell of different height and frequency from the others. In addition, where the ship happens to be a wind may be blowing, so that waves are on the way to becoming fully developed seas. At the same time, the wind may be creating capillary waves on the surface of the larger waves in the area. All in all, the situation looks very complicated.

But we now have at our disposal a method of finding the basic order amid all of this apparent confusion. We know that at any point on the ocean surface we must combine the contribution from each of the waves present, making sure that we take account of whether the wave adds or subtracts from the final result. This process is shown picturially in figure 9-6.

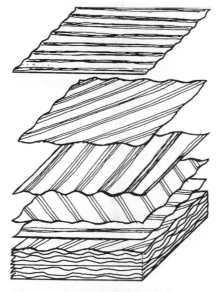

Figure 9-6.

Alternatively, we could represent the addition of these waves through the use of a series of lines on a graph, as shown in figure 9-7. In this

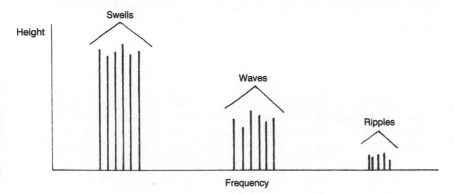

Figure 9-7.

case, the cluster of lines at low frequency (long wavelength) represents the various swells that are the result of distant storms. The lines at intermediate frequency (and therefore intermediate wavelength) represent the waves generated by the winds blowing at the observation point, waves that are not as yet fully developed. Finally, the cluster of lines at short wavelengths (high frequency) corresponds to the capillary ripples caused by the sporadic gusts and breezes striking the water. It is the mutual interference of all these waves taken together that gives us the complex surface we observe in the real ocean.

The Fourier Series

The idea that complex shape can be thought of as the sum of a series of simple wave forms was first introduced into physics by Jean Fourier in 1807. Fourier was a political protégé of Napoleon, and held a series of diplomatic and governmental posts until 1815. After the disappearance of Napoleon from the French political scene, he settled down to a quiet conventional life as an officer of the Academy of Sciences, where he produced a number of mathematical papers on the process by which heat diffuses through solid bodies. His chief claim to fame, however, is not his contribution to the theory of heat but the mathematical technique that he developed almost as a side issue in his studies— a technique known as Fourier analysis. We have already seen the essential feature of Fourier analysis at work in our description of the real ocean wave when we said that combining as few as two different waves could produce a complex surface shape. What Fourier and those who followed him showed was the converse of this statement—that any shape, no matter how complex or irregular, could be produced by combining enough different waves. The sum of the individual waves is known as a Fourier series.

At first glance, this may seem like just another of those interesting but not very useful conclusions that mathematicians prove from time to time. In fact, the Fourier series is a tremendously useful tool in all branches of science. Let me make this point by looking at one practical result related to the ocean waves that have occupied our thoughts. Suppose you were charged with designing facilities in a harbor that was protected from the open sea by a large concrete breakwater. One thing you would want to know is what happened to waves when they encountered the breakwater. Some part of the wave front would be stopped by the concrete, of course, but other parts would enter the harbor through the opening left for boats. Analyzing what will happen in the harbor when the wave hits is not simple, particularly if the incoming wave has the kind of convoluted shape generated by the processes shown in figure 9-7.

The use of Fourier's technique can make this problem much easier to solve. The solution is carried out in three steps. First, you take the incoming wave and break it down into its simple components—in effect, you start at the bottom of the stack in figure 9-6 and work upward. Next, you solve the harbor problem for each of these individual components. Because the wave has been reduced to a simple form, it is much easier to carry out this step than it would be to solve the problem with a real wave. Lastly, after you know how each of the simple waves behaves in the harbor, you add them up again to find the final water configuration of the surface. In effect, Fourier's technique allows you to treat every physical problem involving waves in terms of the simplest possible waves. Dealing with the true complexity of the real world is reduced to the relatively simple task of adding up the simple shapes.

This technique has applications in all situations where waves are known to be present. In TV and radio signals, for example, the amount of each frequency wave that goes into making up the signal is of prime concern to engineers who have to keep different signals from overlapping. Similarly, those who design sound equipment, from musical instruments to stereo systems to auditoriums, have to worry about the behavior of complex sound waves once they get into enclosed spaces.

What makes the technique especially interesting is that it can also be used in situations where the presence of waves is not obvious. For example, engineers often wish to calculate the vibrations that result from winds buffeting a building or a bridge. The force of the wind in a typical storm might look like the graph in figure 9-8. In response to this force, a building might sway and vibrate, and at a given moment might take the shape shown in exaggerated form in figure 9-9. Both

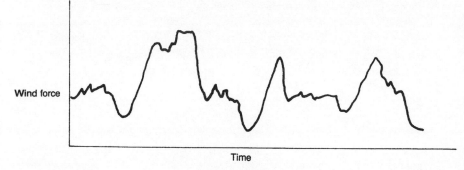

Figure 9-8.

Figure 9-9.

the wind force and the building configuration are complex shapes; but following the lead of our wave-and-harbor example, we know that each of these complex shapes can be broken down into a sum of simple shapes. The problem of dealing with a real building in a gusting wind, therefore, can be replaced by the much simpler problem shown in figure 9-10, in which a wind whose force changes smoothly in time interacts with a building whose shape is mathematically simple.

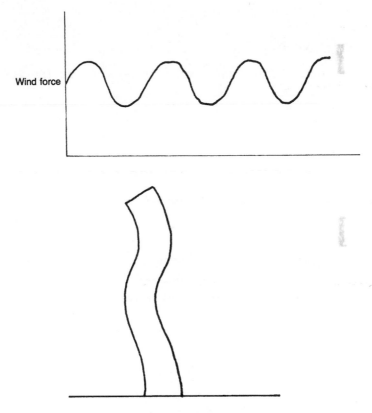

Figure 9-10.

This example could be multiplied endlessly. I speak from personal experience, since I have applied the Fourier series in projects as widely separated as research on artificial heart valves and the design of equipment for fabricating fuel for nuclear reactors. A particularly unexpected application will be dealt with in the next section.

Lest we leave feeling that all problems in physics can be solved by techniques like Fourier's, let me interject a note of caution. The entire system depends on the fact that once we solve the simple problem, we can construct the real solution by adding simple solutions together. In the jargon of physics, a system which has this property is said to be linear. All of the examples we have discussed so far share this property. There are, however, systems in nature that do not—breaking surf is one example. Such systems are said to be non-linear, and until quite recently most problems relating to them simply couldn't be solved.

Paleoclimatology

The last place you might expect to be able to use the kind of wave manipulation we've been discussing is in the analysis of weather. After all, there is nothing "waving" in a set of temperature readings or rainfall indices. This attitude, however, fails to take into account the ubiquity of wavelike phenomena. We know, for example, that temperatures follow a general twenty-four-hour pattern—warm in the day and cold at night. In addition, we know that there is an annual variation in temperatures—it's warmer in summer than in winter. Thus, if we plotted the temperatures at a particular point over a year, we would get a curve like that shown on the left in figure 9-11. The daily fluctuations would take the form of a twenty-four-hour ripple on top of a gradual variation corresponding to the changes of the season. You will notice that from the purely mathematical point of view, this graph could just as well represent a capillary wave produced by a fresh breeze on the surface of a long ocean swell. All we would have to do to make the chance is replace the label on the vertical axis of the graph. Instead of "Temperature," we would write "Water level." Since such a sleight-of-hand could not possibly affect the actual content of the graph, and since the Fourier series deals with mathematical shapes without ever specifying what those shapes belong to we can go ahead and apply the technique we developed for analyzing ocean waves to this "temperature wave."

When we do so, we find a graph like the one shown on the right in figure 9-11. There are two spikes in the graph, indicating that the "wave" was formed by combining waves of two different frequencies—daily and yearly. The spikes will not be perfectly sharp as they were in the simplified water waves we have discussed up to this point. The reason for this is simple. It happens occasionally that the highs and lows of tem-

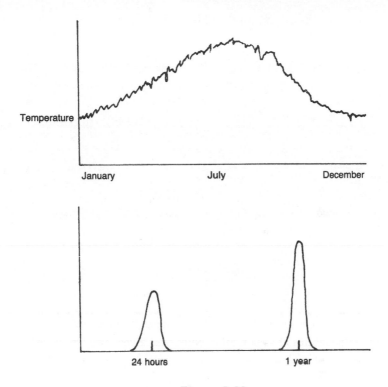

Figure 9-11.

perature on a given day are not exactly twelve hours apart. If a front moves through an area, the night may even be warmer than the preceding day, or the day colder than the night before. This means that the ripple in figure 9-11 is not precisely regular, but has some waves with longer and shorter lengths than the average. This fact is translated in the graph of wave heights as a broadening of the daily peak. Similarly, because some years differ from others in the pattern of their seasonal change, the yearly peak is broadened as well.

Looking at this result, we note one important characteristic: we can indicate the simple physical mechanism that explains the presence of each spike. The fact that the earth turns on its axis every twenty-four hours explains one peak, while the fact that it takes the earth one year to move around the sun explains the other. The question naturally arises whether similarly convincing explanations can be provided if we look at the earth's weather over much longer time scales.

The term "weather" is usually reserved for phenomena that occur over a period of a few years at most, while the term "climate" refers to longer periods. Thus, a hot summer is a change in the weather, while an ice age is a change in climate. The question we have posed, then, is this: Can we provide simple physical explanations for changes in the earth's climate?

The only way to answer this question is to plot the earth's temperature over many thousands of years and then analyze the resulting "wave." But to do this presupposes that we know something of the earth's weather over long periods. This is no modest requirement, since accurate weather records have been kept for only the last few hundred years, and even then only in a few places. As a faculty member at the University of Virginia, I happen to know that the weather records kept by the university's founder, Thomas Jefferson, are an important part of a very sparse data collection in North America. Since the thermometer itself wasn't invented until the seventeenth century, any temperature record going back over three hundred years has to be based on deductions from other information. The story of how old weather records are reconstructed is fascinating, but for our purposes we need know only that it has been done. When a plot of the height of the temperature wave versus frequency is made, we get something like that in figure 9-12.

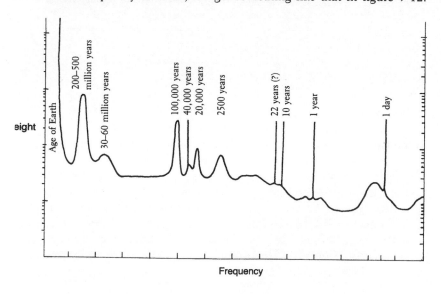

Figure 9-12.

The earth's climatic history shows plainly as a series of sharp spikes against a uniform background.

With one exception, all of these spikes can be explained simply. The daily and yearly spikes we have already discussed. The broad peak associated with time scales of a few weeks has to do with the time it takes for the atmosphere to respond to changes of energy input. The bump at 10 years has to do with the time it takes energy to be transferred back and forth between the atmosphere, the ocean currents, and the earth's ice sheets.

The peak near 20,000 years represents a much longer wave. Its cause has to do with the long-term behavior of the earth's axis of rotation. Right now, that axis points at the North Star, a fact which explains why the North Star appears to be stationary in the sky. The axis does not always point at this particular star, however, but sweeps out a circle in space, somewhat like the axis of rotation of a wobbling, spinning top. Every 23,000 years, the axis makes one complete circuit (as measured relative to the point in the earth's orbit closest to the sun). It is this motion of the earth which is believed to cause periodic ice ages.

The next two spikes also have astronomical explanations. Every 41,000 years, the angle at which the earth's axis tilts undergoes a small change, and every 93,000 years the earth's orbit goes through a regular change of shape due to the presence of the other planets. We would expect both of these effects to produce changes in the weather—an expectation borne out by the presence of the spikes near 40,000 and 100,000 years. Wavelengths longer than this—the peaks corresponding to many millions of years—are thought to be due to geological effects such as continental drift.

In fact, the only spike on the graph surrounded by controversy is the one corresponding to a wave with a period of around twenty years. This is the famous sunspot effect (infamous might actually be a better term as far as some scientists are concerned). I've put a question mark on this spike because part of the sunspot controversy concerns itself with whether there really is a regular twenty-two-year effect in the weather. The other part of the controversy centers on what mechanism could account for the effect if it is there—presumably it has something to do with the interaction of the solar wind with the earth's magnetic field. I mention the sunspot controversy here so that the reader will not be misled by the graph. Scientists do not know everything there is to know about the earth's climate yet.

Having made this point, we still see that the use of Fourier analysis allows us to apportion the causes of the earth's long-term climate among the various meteorological, astronomical, and geological factors that affect it.

Dispersion

By this time, the idea that complex waves can be broken down into simple ones should be well established in your mind, but there is probably a nagging doubt about the reality of the whole operation. The mere fact that some mathematical sleight-of-hand seems to work doesn't imply that the simple waves are really there in nature.

Yet there is something we can see in waves to show that this sort of pessimism is unwarranted. When we introduced the topic of real waves, we noted two important properties: they had complex shapes and those shapes were observed to change with time. The configuration of any wave moving across the ocean will not necessarily be the same at one point in its path as it is in another. You can see this for yourself if you watch the ripples caused by dropping a pebble into a still pond. As the waves move away from the point of impact, they become broader and less steep than they are at the beginning.

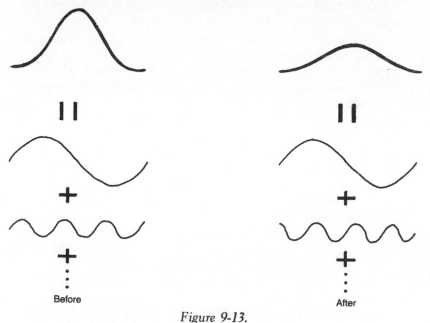

Figure 9-13.

The fundamental reason for the existence of this effect is that simple waves in deep water do not all move at the same speed. In fact, each wavelength moves at a slightly different speed, with longer wavelengths moving faster than shorter ones. In the jargon of physics, we say that waves in water exhibit dispersion. To see why dispersion leads to changes in the shape of moving waves, consider the wave on the left in figure 9-13. A wave shape of this form can, of course, be decomposed into a set of simpler waves, some of which are shown in the lower part of the figure. As the initial wave moves along, the different simple waves which compose it will move at different velocities, so that when we look at the system again (as shown on the right), they will have moved with respect to one another. The delicate balance by which a large number of components added up to a single wave of a specific shape is thus destroyed. The new shape of the wave is obtained by adding up the components in their new relative positions, a process which gives us the smeared-out result shown.

The effects of dispersion can be seen in water waves. They are also well known to communications engineers who have to send signals through the atmosphere. Even if there were no mechanism in the atmosphere for absorbing the energy of electromagnetic waves, engineers would still have to contend with the blurring of their signals due to dispersion. This is particularly important in long-range communications. So the experience of practical men reinforces the conclusions we reach based on our own observations of waves in water: the Fourier components of a complex wave really do exist. They are not simply a mathematical convenience.

10

QUEEN ANNE'S PUDDING

The Walrus and the Carpenter
Were walking close at hand:
They wept like anything to see
Such quantities of sand . . .
"If seven maids with seven mops
Swept it for half a year,
Do you suppose," the Walrus said,
"That they could get it clear?"
"I doubt it," said the Carpenter,
And shed a bitter tear.

—LEWIS CARROLL,
Through the Looking-Glass

I remember being told once that the lawyers at the Inns of Court in London observe an unusual ritual. Every year on a particular date, a large plum pudding is served as part of the banquet. The story is that during the eighteenth century Queen Anne honored her barristers by making them a plum pudding with her own hands. At that time, a small amount of the original pudding was set aside and mixed in with the next year's batch, from which a small amount was again set aside, and so on down to the present. Thus, there is some small but finite probability that part of the plum pudding served to an individual in 1985 was actually handled over two centuries ago by Queen Anne.

I mention this tradition because it bears a very close resemblance to what happens when you pick up a handful of sand at the beach. There is a probability (albeit a small one) that one of the grains in your hand is actually the first grain of sand ever produced on the planet earth. And if you have less chance of getting that first grain than of getting a piece of Queen Anne's handiwork, just remember there's a lot more sand in the world than plum pudding.

You can start your investigation of sand by looking at your handful for a while. You might think there could be nothing so alike as two grains of sand on a beach, but if you look at what you're holding you'll quickly see that the grains are not identical. Here and there in a handful of light-colored sand, for example, you will see a grain of darker material. Occasionally, sunlight will glint off a small piece of mica. If you looked through a microscope, even more differences would become apparent. Some grains would look sharp and angular, as if they had just been chipped from the living rock; others would look smooth and rounded, with opaque frosty surfaces that do not reflect the light evenly. All of these differences are important, for the sand you are holding comes from many different sources and has had different histories. It is not an exaggeration to say that each grain has its own unique story to tell, a story we can often unravel by a judicious application of elementary physics.

From the point of view of a geologist, sand is simply an accumulation of particles whose largest dimension is between .05 and 2 millimeters. A dime is roughly a millimeter in thickness, so a collection of particles whose diameter is the thickness of two dimes would be classified as sand (albeit a rather coarse one). So would a collection whose particles had diameters less than one-tenth the thickness of a single coin. Particles with diameters greater than 2 mm are considered to be gravel, those less than .05 mm are silt, or, if the diamaters are less than .004 mm, clay. It isn't necessary that the particles have any particular composition, which explains why we use the term "sand" to describe the black material on some Pacific beaches as well as the white stuff found elsewhere. Any material can be made into sand. The only requirement is that the particles be the right size. The most common sand—the one you're most likely to encounter at your favorite beach—is derived from quartz. Chemically, it consists of one molecule of silicon and two of oxygen, a combination known as silicon dioxide.

There is actually a lot more sand in the world than you might think when you look at an ocean beach, where sand seems to form a marginal ribbon between land and sea. Most of the sand in the world is not located along seashores. Some of it exists on the surface in deserts around the globe, and still more in underground deposits, where it is sometimes mined for the traces of rare minerals it contains. The lowest estimate of the amount of sand on the earth puts it at around 10 million cubic miles—enough to cover the entire United States with a layer three miles thick. Almost all of this sand is located on the continents and those parts of the ocean immediately adjacent to continents. There is very little sand in the deeper parts of the sea. This fact gives us our first hint as to where the sand comes from. It tells us that sand is created from dry land, so that is where we should begin our search for its origins.

From its chemical composition, we know that sand must have come from continental rocks in the process of being weathered. It used to be thought that as these rocks were carried downstream by rivers and brooks, they would be worn down until all that was left was a grain of sand and some finer material scattered along the path the rock had taken. In this view, the sand we see would represent a rock in the last stages of being returned to dust, and no grain of sand could be expected to stay around very long. Thus, each grain of sand represents the remains of a large rock, the rest of the rock having been deposited as dust along the streambed. But a grain of sand is only a millionth the volume of an average rock, so this process would have to produce a trainload of silt and clay for every shovelful of sand at your beach. There is just too much sand in the world for it to have been produced by this sort of process.

Sand-sized grains can be produced in many ways, including chemical reactions and formation by living organisms. The dominant mode of production, however, seems to be the weathering of rocks which have been uplifted in the process of mountain formation. If you want to look for the ultimate origins of your favorite beach, then, you have to travel to distant mountains. There the forces of weather, vegetation, and gravity combine to break rocks down into small pieces, strewing the high mountain valleys with debris consisting of every size of rock from silt up through sand and gravel to boulders the size of a house. Rain and snow runoff start this debris on its way to the ocean. It isn't only the sand and gravel that can be transported this way, either. I can vividly remember trying to go to sleep next to a stream in Montana, all the while listening to the booming sound made by barrel-sized boulders being rolled downstream in the spring flooding.

As the streams leave the mountains, the size of particles that can be transported by the water decreases to match the newly moderated rate of flow. As the streams flow into rivers traversing relatively level flood plains, heavier particles continue to drop out. At the present time, rivers in North America are moving so slowly by the time they reach the sea that they carry only silt out onto the continental shelf. Ten thousand years ago, however, when the great glacial sheets were melting and the level of the oceans was lower than it is now, it was otherwise. Large amounts of sand were carried to the sea at that time by the runoff, and perhaps some of the grains in your handful reached the shore when the Neanderthals were still roaming northern Europe.

This picture of the movement of sand to the sea explains many things about your handful of sand. Since large rivers are fed by numerous streams, it is only reasonable that grains generated from many different formations will be mingled together as each tributary makes its own contribution to the total. This also explains why it was that people thought that sand was made from worn-down boulders, for if you follow any given stream from the ocean back to its source, you will find it transporting larger and larger bits of material.

It would be wrong to picture the progression of a grain of sand from an alpine valley to a beach as a smooth, uninterrupted transit. In fact, grains do not move continuously at all. A given grain may start down a mountainside, get trapped in the roots of a plant and remain in one place for years until a spring flood washes it out, move down to the plain to be trapped behind a large boulder, stay there for a while, and so on. Indeed, if the sand happens to be deposited high on the banks of a river or out on the plain during a flood, it may be thousands of years before it resumes its journey to the sea. As a rule of thumb, the average river requires a million years to move a grain of sand one hundred miles closer to its destination. Given this leisurely time scale, the ten thousand years since the last ice age really doesn't amount to much as far as the overall movement of sand is concerned.

Once the sand does reach the seashore, any number of things can happen. It can spend some time as part of a beach or sand dune, being blown around by the wind or moved back and forth by the currents. It can also be carried out from the shore and deposited on the ocean bottom. Since storm disturbances rarely go down more than a hundred feet below the surface, the grain which is deposited on the continental shelf is likely to remain in place, eventually being buried as more sand falls down on top of it. Occasionally, sand deposited near the edge of the continental shelf will slide off into the deeper ocean floor in what geologists call a turbidity current, but whether it winds up on the shelf

against your ankles. You can go down to the waterline and look at the wash running up the beach. If there's not too much foam, you can see the cloud of sand grains suspended in and moving with the water. Sometimes, if the surf is particularly rough, you can see great streaks of suspended sand in the roiling waves themselves. Finally, the entire wave-sand system moves up and down the beach twice a day as the tides come in and out. And as we shall see shortly, forces are at work moving the sand along the face of the beach as well as up and down with the waves.

By far the chief mechanism for moving sand about is the process called entrainment, by which the grains are picked up and carried along with moving water. The question of whether a particular grain will be picked up is complex, depending on the speed of the water, whether the flow is turbulent or not, how heavy the grain is, what shape it is, and what sort of bottom it is lying on when the water arrives. For our present purposes, we need only make a few common-sense observations: swiftly flowing, turbulent water will pick up grains more easily than a

Beaches need not always be sand. This is a beach made from pebbles at Bean Hollow State Beach, California.

flow that is less swift or turbulent; and it is easier for a given flow to pick up light grains than heavy ones.

In chapter 7 we saw that as a wave moves in toward shore, the bottom begins to affect the movement of the water when the depth becomes comparable to the wavelength. At depths greater than this, the passage of the wave has very little effect on the motion of the water at the bottom, and hence little effect on the sand which finds itself there. As the depth decreases, however, the water at the bottom executes a regular back and forth motion, which becomes more violent as the wave begins to crest and break. After the wave breaks, the water runs up the beach, slowing down as it does so. Eventually the wave stops, some water sinks into the sand, and some runs back down to rejoin the sea. At most points on the bottom, then, the water executes a closed cyclical motion, first moving toward the beach and then away. There is, on the average, no net motion associated with this sort of cycle.

Yet this fact does not imply that there can be no net motion of the individual sand grains. To give one example, on a steeply sloping beach a dislodged grain may roll down the slope because of gravity. The grain will not, of course, move up the steep slope very quickly when the water is going in that direction. In this example, the effect of gravity is to produce a net motion of sand grains away from the beach. On the other hand, the configuration of another beach may be such that water carries sand along with it on the way in and sets the sand down before starting back, producing a net motion of sand toward the beach. Many situations other than these two result in a net motion of sand, and so many factors come into play that it is difficult to predict which way the sand will move in a given situation.

One reason for this uncertainty is that each wave does not have to move a sand grain very far to achieve striking sand transport effects. For example, if the waves at your favorite beach come in every ten seconds, there will be $10 \times 60 \times 24 = 14,400$ waves per day. If each wave moves an individual sand grain only 1/16th of an inch, the net movement in a day will be 900 inches, or 75 feet. Thus, it wouldn't take long for that grain to migrate far from its original position.

There is one general rule of sand motion that oceanographers have noted amid all the complexity. It appears that waves whose height is small compared to the wavelength—"small" waves—tend to move sand shoreward, while waves whose height is larger move it away from shore. Roughly speaking, if the ratio of wave height to wavelength is less than .03 (e.g., if a wave 100 feet long is less than 3 feet high), the wave is "small." If the ratio is larger than .03, it is "large." The mechanism for providing shoreward motion seems to be the one given in the examples above: sand is lifted up when it moves toward shore and drags along the bottom on the return. When the waves are larger—as they might be during a storm—the general turbulence of the surf increases and the sand stays in suspension long enough to be returned to the place where the breakers form. At this point the sand is deposited and an underwater sandbar is formed.

Since severe storms and large waves are likely to occur in the winter, while mild seas occur in the summertime, this analysis of sand transport by waves suggests that there should be a seasonal shift in the shape of beaches. In the summer the sand should be piled up on the shoreline, while in the winter the waves should remove the sand from the beach and deposit it in an offshore bar. This general pattern is, of course, what we find on many beaches. The seasonal motion of sand is superimposed on any long-term processes of erosion or accretion which may take place at a particular beach.

An example of beaches protected by rocky headlands, and, therefore, effectively isolated from the littoral conveyor belt. Bean Hollow State Beach, California.

Marks made by runoff from a beach—the same runoff that is part of the littoral conveyor belt.

If the beach were completely isolated, so that no sand could be moved away from the surf zone, the seasonal shift would be the only major change. In some beaches located in coves between headlands this sort of isolation is almost achieved, and much of our knowledge of beach mechanics comes from studying such simple systems. However, many beaches consist of stretches of sand many miles long, so that in addition to worrying about sand transport toward and away from the shore, we have to detect the motion of sand up and down the coast as well.

A number of mechanisms move grains laterally along the beach. In chapter 7 we saw that the refraction of ocean waves causes surf to wheel around and approach the beach in a more or less perpendicular direction. If the shoreline were perfectly straight and the bottom absolutely uniform, the surf would approach exactly along the perpendicular. In the real world, this situation is rarely met. Usually the surf approaches the shore at a small angle, determined by the details of local geography. When the waves approach a beach in this way, the water moves up the beach at an angle, stops, and then flows straight down the slope under the influence of gravity. A sand grain suspended in the water will therefore be carried up the beach in an angular direction and down a straight line. The next wave will again move it up the beach at an angle, and it's not hard to see that the effect of a series of waves will be to move the grain down the beach in a sawtooth pattern.

In addition to this rather straightforward effect at the beach front, the shoreward or seaward motion of bottom sand due to the waves will also be at an angle to the beach. Thus, if the incoming waves are small, they will induce a net motion of sand at an angle toward the beach. If this happens up and down a long coast, the effect can be thought of as the sum of two separate motions: the usual motion toward the shore, and an additional small motion along the shore. Both the sawtooth motion of grains at the beach and the angular motion of grains on the bottom result in sand being moved in a direction parallel to the beach.

Willard Bascomb, in his excellent book *Waves and Beaches,** describes the motion of sand along a beach as "the littoral conveyor belt." He points out that oceanographers think of the movement of sand as being like a huge excavating operation. The waves play the role of the bulldozers, digging the sand up, and the phenomena we've just described play the role of a conveyor belt, moving the sand sideways along the beach.

Since there is a limited amount of sand on a given beach, the presence of the littoral conveyor belt has enormous consequences for people and communities who own property along the sea. For unlike other kinds of real estate, the boundaries of such property are likely to change with time. Depending on where you are located with respect to the littoral conveyor belt, you may find that your beach is getting larger as sand accumulates, or you may find your beach, your shoreline, and even your house being eaten up by the advancing ocean. People who have paid tens or hundreds of thousands of dollars for a seaside building lot are likely to demand that local government step in at this point and do something.

One way to stop the sand from leaving a particular area is to build a long wall (called a groin) perpendicular to the beach. The groin will trap sand moving down the littoral conveyor belt, piling it up and creating a new section of beach. Unfortunately, this particular solution means that properties on the far side of the groin will no longer receive the supplies of sand they once did, and beaches there will start to erode as the interrupted conveyor belt starts up again. A classic example of this can be seen on the Atlantic coast a few miles east of Washington, D.C., in the Maryland town of Ocean City. In 1953, faced with the

*New York: Doubleday, 1980.

The movement of sand on a beach is well illustrated by this buried railing at Assateague National Seashore, Virginia. The steps themselves are three feet below the surface.

erosion of beaches and the consequent loss of tourist and real estate revenue, the local government constructed several large groins and jetties to trap sand. The operation was successful, and in the intervening years developers in Ocean City have made fortunes building beachfront homes and condominiums. At the same time, Asateague National Seashore, an undeveloped stretch of beach directly to the south, has been deprived of its normal supply of sand. The result: the northern tip of the island has disappeared and five miles of beach have become severely eroded.

The littoral movement of sand isn't the only problem facing property owners on the Atlantic coast; the level of the ocean is rising at the rate of eight to ten inches per century. Given the low level of the eastern seaboard, even a small increase in average water level can cause significant shifts in the land areas that can be reached by storm waves. In a typical year, for example, about 150 houses are destroyed or severely damaged in the Outer Banks of North Carolina alone.

As the waves begin to encroach on buildings, the first defense is usually the construction of a sea wall—a heavy concrete or rock wall

The erosion of dunes is very clear in this photograph, taken a few days after Tropical Storm Dean came ashore at Assateague National Seashore in October 1983.

Further damage from Tropical Storm Dean at Assateague National Seashore. The area on the right, beyond the dune remnants, is a paved road now under more than a foot of sand.

designed to hold the waves back. This defense can work for a while, but we know enough about the seasonal movement of sand to be sure that it won't last for long. At first, the sea wall will be reached only by storm waves—large waves that carry sand away from the beach. In the absence of the wall these waves would remove sand, but would leave a gently sloping beach; in effect, they would move the beach inland. With the wall present they still remove sand, but cannot run up on the shore. Consequently, after the storm the bottom has a much steeper slope than it had before. This means that incoming waves will crest more quickly, and that fewer small waves will come to the wall. Since under normal circumstances it is small waves that return the sand to the beach, the net effect of erecting a sea wall will be to protect the buildings behind it, but only at the cost of destroying the beach in front of it. As the house was probably built because of the beach in the first place, the whole system is clearly self-defeating.

A classic case of sea wall–beach interaction can be seen in Miami Beach, Florida. When sea walls were built to protect the expensive waterfront hotels, the beaches that gave the city its name soon disappeared. In 1981 an extensive rebuilding project was finished, complete with new sand dredged from the ocean. The cost: $4.5 million per mile ($62 million dollars by 1992), and the sand will probably have to be replenished every ten years or so. The cost of interfering with nature is clearly very high.

If attempts to preserve the present coastline inevitably lead to destruction, if we can keep our beach in operation only by beggaring our neighbor's, what can be done about coastline property? For some very high priced land, such as that at Miami Beach, the answer is clear: We rebuild the beach, no matter what it takes. In other places, the federal government has already stopped issuing low-cost storm insurance, a move which will effectively prevent further building. In these places, the natural beach processes will be allowed to continue and the shoreline will evolve. There really doesn't seem to be a reasonable middle ground. In the words of Orrin Pilkey of Duke University, who has done as much as anyone to alert the public to what is happening to American beaches, "We have two mutually exclusive choices; beaches or buildings. We can't have both."

11

OF BEACHES
AND BASEBALLS

Close by the sturdy batsman the ball unheeded sped
"That ain't my style," said Casey; "Strike one," the umpire
said.

—E.L. THAYER,
"Casey at the Bat"

The great insight that physics more than any other discipline gives us is that once we have deduced a general law of nature it must apply to every phenomenon we observe, no matter how different those phenomena may appear on the surface. The laws that govern the movement of sand on the beach, the waves washing up on the shore, the tides that govern the water level, and even the orbit of the moon that raises those tides are the same. That they can be applied to the greatest astronomical body and the smallest grain of sand is an indication of the breadth of the insights that went into formulating the laws in the first place.

The first formulation of universal laws was the work of Isaac Newton.

In his *Principia Mathematica*, published in 1687, he laid down three principles which govern the motion of all material objects. The principles, now known as Newton's Laws of Motion, are quite simple:

Law 1. An object remains at rest or in a state of uniform motion unless acted on by a force.

Law 2. The reaction of the body as measured by its acceleration is proportional to the force applied and inversely proportional to the mass.

Law 3. For every action there is an equal and opposite reaction.

Everything in the world that moves (or remains unmoving) does so according to the rules laid out in these simple statements.

Before we look at how these principles make themselves manifest on the beach, we should probably stop to contemplate the fact that, innocuous as they seem to us today, in their time they amounted to a revolution in scientific thought every bit as great as that brought about by relativity and quantum mechanics in the early years of the twentieth century. For example, the first law says that an object will move in a straight line (or remain at rest) unless some outside agency exerts a force on it. In other words, "natural" motion—motion associated with no outside interference—is rectilinear. This statement was a direct contradiction of what had been believed by scientists from Aristotle until well after the time of Copernicus, all of whom had accepted without question the statement that an object left to itself would move in a circular path, the circle being the most perfect geometrical figure.

This preoccupation with the circle led to all sorts of difficulties in pre-Newtonian science. For example, since heavenly bodies were thought to be free from all external influences, they were required to move in circular paths. The observed complexity of the motion of the planets (a complexity we now understand to be due in part to the fact that their orbits are elliptical) led to cosmologies in which the planets moved on circles rolling on circles rolling on circles. The name for a circle rolling on another is *epicycle*, and some models posited more than seventy epicycles in the solar system. The complexity of the classical vision caused Alfonso the wise, king of Castille in the thirteenth century, to remark that "If the Good Lord had consulted me before embarking upon creation, I would have recommended something simpler." He couldn't have known, of course, that God had actually created something simpler, and that the motion of the planets would someday be understood in terms of three laws of motion and the law of gravitation (see chapter 3).

The primary force acting on a planet is the gravitational attraction of the sun. If this force were suddenly turned off, the planets would move away from the solar system following a straight-line course (as indicated in the first law). We see, then, that the actual orbits of the planets are highly "unnatural," in that a constant force is required to maintain them. In understanding the orbit of a planet, as in any other problem involving motion, the first question that must be answered is: What forces are acting? This question is relatively easy to answer for planets.

For a grain of sand resting on the bottom of the sea or on a dune, the situation is more complicated. The principles are the same—Newton's laws still apply—but the constellation of forces is much more complex. In figure 11-1 we show such a grain, together with all the forces acting on it. You will note that the forces do not all act in the same direction.

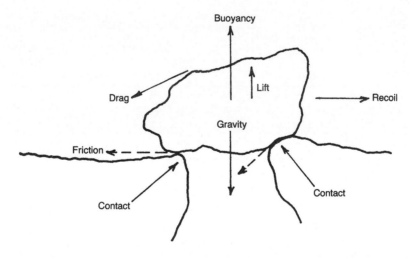

Figure 11-1.

The easiest force to understand is gravity. Acting on the grain, it tends to pull it down toward the center of the earth. This force will always be present, no matter where the grain is located. Next, there will be forces associated with the points of contact the grain has with its neighbors. One of these will be the ordinary contact force that exists between any two bodies which touch each other—it's the force you feel with the soles of your feet when you stand up or on the palm of your hand when you lean against a wall. This force is directed perpendicular to the plane of contact between the grain and its neighbors.

The second force being exerted through contact is that of friction. This force is shown by a dotted line in the figure because it does not have a constant direction, but instead always acts in such a way as to oppose the direction of motion. Thus, if the grain is being pushed to the left, the frictional force will be acting to the right, and vice versa. You are familiar with this force in other situations. For example, if you are trying to push a heavy piece of furniture across the room, you have to exert quite a bit of force in order to get it moving. It is friction that is opposing the force your muscles are exerting.

Furniture

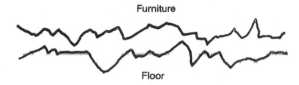

Floor

Figure 11-2.

If you could look at the interface between the furniture and the floor with a microscope, you would see something like what is shown in figure 11-2. The apparently smooth surfaces, when examined under magnification, turn out to be rough and uneven. In order to move the upper surface, we have to exert enough force to break off some of the protrusions on both surfaces. Once we do this, they will slide over each other more easily—a fact which explains why it is easier to keep something moving than to get it started in the first place. Thus, we see that there will always be forces acting on the grain of our grain of sand that will tend to keep it from moving.

Up to this point, we have discussed only those forces that are associated with the weight of the sand grain and its interaction with its neighbors. If we were talking about a grain of the surface of the moon, this would be the end of the story. None of the forces yet enumerated could cause motion, so the grain would just stay where it was. On the earth, however, the grain will find itself immersed in a fluid—either water (if it's in the ocean) or air (if it's on a dune). The presence of the fluid brings a number of new forces into play.

If the fluid isn't moving, then the only force it exerts is pressure. The mechanism involved is the collisions of fluid molecules with the surfaces of the sand grain. Each collision results in a small force being exerted on the grain, with the accumulation of these individual forces resulting in the sort of thing you can measure with a tire gauge. So long as the water or air is stationary, this collision-induced pressure will be exerted on all sides of the grain. Because the bottom of the grain is at a slightly greater depth than the top, the upward pressure exerted on the bottom will be slightly greater than the downward pressure exerted on the top. The result is a net upward force on the grain, a force which counters the downward force of gravity. This imbalance in the pressures is actually nothing more than the familiar buoyant force that keeps us afloat when we go swimming. Its net effect on the sand grain is such that the grain weighs less when it is submerged than when it is not. The buoyant forces are particularly important when the grain is in water, for they tend to keep the grain from settling back to the bottom once it has been lifted up.

If the fluid is moving, as it would be if the grain were on a dune exposed to the wind or a bottom exposed to wave motion, a number of new fluid forces have to be taken into account. If you've ever used a hose to clean a sidewalk or driveway, you have direct experience with one of them. When a stream of water is directed at an object, the object will recoil in much the same way as it would were the water replaced by a stream of particles. The water bounces one way, the object moves the other, as required by the action–reaction principle of the third law. In the case of your cleaning operation, the dirt recoils off the cement while the water bounces up into the air. In just the same way, there will be a force on the sand grain arising from the mass motion of the water. Depending on which way the water is moving and the slope of the bottom, the grain can be pushed up or down the slope or even be jammed more tightly against its neighbors. In figure 11-1 we have shown this recoil force being directed downward.

A fluid moving across a solid surface tends to drag the surface with it, an effect known as viscous drag. The most common example of this effect is the resistance of the wind to a moving automobile, a phenomenon due in part to the drag of the air moving across the surface. Yachtsmen also know about viscous drag, although in their case the force is more likely to be exerted by water on the hull of the boat. From the point of view of the physicist, it makes no difference whether we are talking about solid objects moving through a stationary fluid (as in these examples) or a fluid moving past a solid body (as in the case of the sand grain). The viscous force will still be there, and it will tend to make the object move with the same speed as the fluid. For the car and boat, this means that it will try to slow the object down: for the sand grain, it means that it will try to pick it up.

The final effect of the moving fluid results in the most interesting force of all—lift. At the top in figure 11-3 we show an idealized sand grain and the fluid flowing around it. In order that there be no discontinuity in the flow, the fluid going the long way around over the top of the grain will have to move more quickly than the fluid which

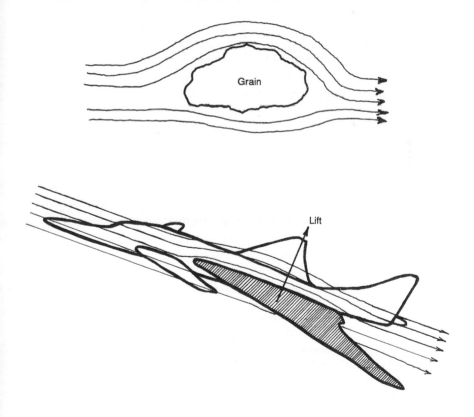

Figure 11-3.

flows under the bottom. Consequently, the existence of the grain causes a certain asymmetry in the flow of the fluid.

This asymmetry, in turn, gives rise to a force on the grain. The best way to understand this force is to think of a number of examples: when a fast wind blows by a chimney, air is pulled up into the flue, an effect you can easily observe sitting next to a fireplace on a stormy night. If a window in a speeding car is opened, loose bits of paper from inside the car are often pulled out—an unintentional bit of littering. In both of these cases, something moves in response to a rapid motion of air elsewhere. Physicists know this as the Bernoulli effect. In essence, if we start with Newton's laws and apply some mathematical reasoning, we find that those parts of a fluid which are moving quickly will exert less pressure on their surroundings than those which are moving more slowly. In the two examples, the pressure at the top of the chimney and at the car window was reduced because of the rapid motion of air. The result was a pressure imbalance on objects inside the house or car, an imbalance which ultimately led to motion.

Turning back to our sand grain, we see that here, too, there will be a pressure imbalance due to the unequal motion of the fluid. The pressure above the grain will be lower than the pressure below, and the result will be a net upward force on the grain. This is the force we have called lift. Whenever you have flown in an airplane or sailed in a sailboat, you have made extensive use of the phenomenon of lift, although you may not be aware of it.

At the bottom of figure 11-3 above we show a typical pattern of wind airflow around the wing of an airplane. From the discussion of lift on a sand grain, we recognize immediately that there will be an upward force on the wing. If the airplane is moving fast enough, this force may even become large enough to overcome the downward pull of gravity. When this happens, of course, Newton's laws tell us that the plane will start to move in the direction of the lift force. Sitting inside, you will notice that the plane is no longer touching the ground, but has started its ascent into the sky. The important point to note here is that it's the relative velocity between the air and the wing that matters, not whether the plane is moving past the air or vice versa. Indeed, in violent windstorms it often happens that small aircraft are lifted off the ground even though they are not moving at all. In just the same way, when water or air moves past a sand grain or the sail of a boat quickly enough, the sail will move or the grain will be lifted.

Before leaving the Bernoulli effect, I'd like to point out one more area where its consequences should be explored, and that is the some-what unexpected activity of baseball. Consider, if you will, the curve ball. This particular pitch is thrown so that the ball spins around an axis as it moves forward, as shown at the top in figure 11-4. Because the surface of the ball is rough, the effect of viscous forces is to create a thin layer of air which rotates with the surface. Looking at the diagram, we see that the air at the point labeled A will be moving faster than the air at the point labeled B, because in the first case the motion of the ball's surface is added to the ball's overall velocity, while in the second it is subtracted. The effect, then, is a "lift" force, which tends to move the ball in the direction shown. From the point of view of the batter the ball curves as it approaches, the direction of the curve being determined by the direction of spin.

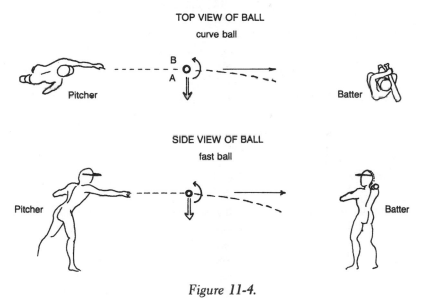

Figure 11-4.

A fast ball on the other hand, is thrown so that the spin is as shown at the bottom in the figure. This spin is imparted by having the pitcher hold the ball so that his fingers are along the laces. By an argument exactly analogous to that given for the curve, we conclude that there will be a "lift" force on the fast ball, which causes it to move downward while in flight. In this case, the batter will see the ball sink as it approaches the plate.

If the baseball were a perfect sphere, this is the only kind of fast ball that could be thrown. But the baseball is not a perfect sphere since it has raised seams on its surface. These seams produce aerodynamic effects which are considerably more complicated than any we have discussed up to this point. Suffice it to say that if the pitcher throws a fast ball with his fingers held across the laces rather than along them, the forces associated with the seams will not only overcome the downward force associated with the Bernoulli effect, but will produce a net upward force. This results in what is known as a rising fast ball. Indeed, when I was younger and blessed with reflexes much faster than those to which, alas, I have lately become accustomed, it was generally agreed that the amount of "hop" on a pitcher's fast ball was the best index of the amount of "stuff" he could put into his pitches.

With lift we have completed the roster of forces that can act on a grain of sand subjected to the influence of a moving fluid. The catalogue illustrates an important point about the way physics works: the general principles may be very simple—even deceptively so—but when we start to apply them to a real-life situation the underlying simplicity is quickly obscured by the large number of factors that have to be taken into account. In the end, though, the analysis of the forces on a sand grain come down to one simple question: Is the net force zero or not? If it is zero (for example, if the grain is being pushed down into the ocean bed), no motion will result. This follows from the first law. (Note that in this case the downward force of the water is canceled by the upward force associated with the contact points or with friction.) If the net force is not zero, then a grain will move, and the motion will be that described by the second law.

Movement in the Air

Let's start our analysis with a grain of sand lying above the high tide mark on the beach. When the wind blows fast enough, this grain will be picked up. Once aloft, however, there is very little buoyancy, since sand is much more dense than air. Consequently, the grain will soon fall back to the ground. The progress of a single grain along the beach can therefore be thought of as a series of short leaps—a process known as *saltation*.

Given this fact, we can see that after the wind has been blowing over the sand for a while, an equilibrium will establish itself. Whenever a grain is deposited in a particular area, another will be picked up, so that although the patch of sand may be moved in the direction of the wind, there is no accumulation at any particular point, provided that the patch was level at the beginning.

But how likely is it that a patch of sand at the beach will be absolutely level? We would probably expect the surface to have a random assortment of hills and valleys in it, since truly level surfaces are a rarity in

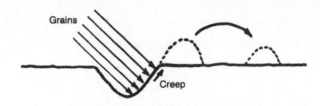

Figure 11-5.

nature. For the sake of argument, let's assume that at one point there is a narrow valley in the surface, as shown in figure 11-5. The saltating sand grains will rain down on the surface at an angle, as shown. This has two effects: first, fewer grains will fall in the sheltered lee side on the valley—a part of the surface that is effectively "in shadow." Second, as grains rain down on the exposed side of the valley, the collisions will tend to push those grains already there up the slope, creating a small hill (shown by a dotted line in the figure). We say that the sand creeps up the slope. Thus, the process of grain saltation will take any preexisting valley and deepen it, producing a hill on the downwind side.

The changes in the surface do not stop at this point, however. The wind is still blowing, and sand grains on the newly created hill will be picked up. If the grains of sand in a patch are of roughly the same size, then the distance a given grain can be carried by the wind will be about the same for all grains. Thus, grains blown off the crest of the hill will travel a characteristic distance before they drop back to the ground, a distance that depends on wind velocity and grain size. The effect is that a certain distance downwind from the original hill a second hill will begin to form, as shown on the right in figure 11-5. The natural pro-

Wind-driven ripples forming on the side of a dune. The wind was blowing parallel to the shore and the ripples formed in less than an hour. Assateague National Seashore, Virginia.

cesses of grain movement will create a series of ripples on the sand.

The ripples will continue to grow by this process until they are high enough to alter the flow of wind near the ground. Basically, as the ripple gets taller, the velocity of the wind at its crest becomes greater; this, in turn, means that more grains get blown off or pushed into the sheltered hollows. Eventually an equilibrium is reached where as many grains are added to the surface as are removed, and the ripple shape becomes stable. In the case of a beach, there is a constant supply of sand brought up to the beach by the waves and currents. After it has dried out, it becomes susceptible to further movement by the wind. Thus, the process of saltation and impact we have described can go on at any time.

From this account, it is clear that the main process at work in building wind-driven ripples is the effect of the impact of falling sand grains on the original sandbed. There is another impact-driven phenomenon that you may have had called to your attention while driving. I refer to the washboard road, an all too common occurrence on unpaved country

Large-scale wind-formed ripples in the sand at Año Neuvo State Park, California.

back roads. In this case it is car wheels which provide the impact that turns a small depression into a hill, and the bouncing of the car which creates the next depression—a depression that will eventually develop into a hill which, in turn, creates another depression downstream. The process continues until a hard spot in the road surface is reached.

Movement in the Water

There is another sort of ripple seen at the beach, one which at first glance seems identical to the wind-caused ripples we have just described. These ripples generally can be seen at low tide and are clearly associated with areas where waves have been coming in toward the beach. Despite appearances, however, the mechanisms leading to beach ripples are quite different from those leading to ripples in dry sand.

The basic difference has to do with the differences between the buoyant forces on a sand grain in air and in water. In the former case, as we have pointed out, there is very little upward force on the sand owing to the surrounding air. This means that when the grain falls it makes a heavy impact on the sandbed, an impact which leads to the movement of grains and the formation of hills. A sand grain in water, however, is acted on by quite a large buoyant force. Indeed, the buoyant force is only a little smaller than the force of gravity. One way to convince yourself of this is to put some sand in the bottom of a glass of water and shake it. It takes a long time for the sand to return to the bottom, a fact which we can take to mean that the normal force of gravity is being largely canceled out. This slow settling means that a grain will undergo a "soft" landing, with little impact transferred to its new neighbors. The movement of grains in water, then, must be governed primarily by the flow of the fluid itself, rather than by the collisons of the sand grains with one another.

Having made this point, we should note that the process by which ripples are formed by moving water is still a subject of debate among hydrologists. Almost every journal article I read on the subject had some comment like "This is what I think causes the ripples, but authors X, Y, and Z each have their own theories." The situation is complicated by the fact that the mechanism which causes ripples to form on river bottoms, where the water flows in one direction only, seems to be quite different from the mechanism which causes ripples to form at the beach, where the waves move the sand back and forth. And both of these mechanisms, of course, are different from those which form ripples in wind-driven sand. I have to admit that this came as quite a surprise to me, because when I saw the ripples on the dunes at Año Neuvo State Park in California (those in the photo on p. 153), I immediately jumped to the conclusion that I had found a beautiful example of the same mechanism working in air as was working in water. It was one of those moments of insight that the French call "the false, clear idea." In this case, nature seems to have chosen to produce similar end results by quite different means in different situations.

The formation of ripples in the presence of moving water appears to be intimately related to the presence of turbulence in the flow. When water flows slowly and placidly, the flow is said to be laminar. At higher speeds, irregularities and eddies start to appear, and we say that the flow is becoming turbulent. If you watch the water running from your kitchen tap, you can see this transition for yourself. When you turn the tap so that the flow is very slow, the water coming out has a smooth surface— you can actually see your reflection in it if you look. As you open the tap wider, ripples and irregularities start to appear until, finally, the jet looks milky because light is reflecting off of all the trapped air bubbles. This is fully developed turbulence.

From the point of view of the development of ripples, the key aspect of turbulent flow is the presence of eddies. In a moving stream, for example, the presence of a ridge on the bottom causes an eddy to form in the flow on the downstream side, as shown on the top in figure 11-6. The flow removes sand from the top and downstream face of the ridge and deposits it at the point labeled A, starting a new ridge. This process continues until the flow pushes as much sand over the top of the ridge as is excavated by the eddy. The result is a series of characteristically shaped ripples as shown on the bottom.

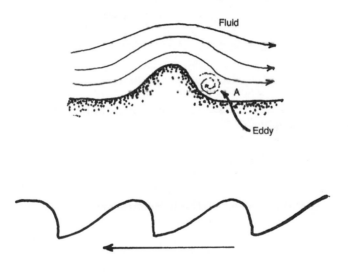

Figure 11-6.

The eddies caused by waves moving near a beach act in a slightly different way, reflecting in part the fact that the motion is usually considerably more turbulent in the surf than it is in a river. If there is a ridge on the bottom, an eddy will form as shown on the left in figure

11-7. The eddy will pick up sand and keep it suspended in much the same way you keep sugar suspended in your tea by stirring. The eddy will have a characteristic time during which it will dissipate, just as the tea will eventually stop swirling after you remove your spoon. When the eddy disappears, the suspended sand settles out. If the return sweep occurs before the eddy dissipates, the excavated sand may be dumped on top of its own ridge. Otherwise, it may be dumped farther along, starting a new ridge. In any case, the result is a set of uniform ripples in the sand—ripples which can usually be seen quite easily at low tide.

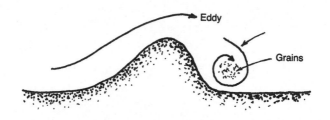

Figure 11-7.

What makes ripple formation such a difficult problem is that the moving water changes the shape of the bottom, but the changed bottom shape then changes the fluid flow. The mathematics involved renders this sort of complicated non-linear feedback problem hard to solve, a fact which explains the current unsatisfactory state of the theory. In addition, at a real beach the two mechanisms outlined above are likely to be operating at the same time, for as well as the waves there are usually currents running along the beach—currents whose behavior is much like that of a river. Consequently, the resulting ripple formations are likely to be some complicated mixture of the two types of "pure" ripples shown in the figures.

The key point in understanding the formations of ripples by moving water, then, seems to be the fact that the buoyant force on sand grains is so strong that the grains settle gently to earth when the fluid flow is slow enough. There is another system which has these same characteristics, and which therefore ought to produce the same sort of ripples as water. That system is fluffy snowflakes in a light wind. All of the arguments we made about sand ripples in a river ought to apply to this situation. And indeed, ripples much like those seen in rivers are often seen on snow-covered surfaces after the wind has been blowing for a while. This is one case, at least, where we can see the same mechanism working in very different situations.

12

CASTLES IN
THE SAND

It happened one day . . . I was exceedingly surprised with the print of a man's naked foot on the shore, which was very plain to be seen in the sand.

—DANIEL DEFOE,
Robinson Crusoe

If you watch your footprints as you walk along the beach, you will notice that they aren't all the same. Up near the dunes, where the sand is bone dry, your trail will be marked by a series of depressions. There will be no detailed imprint of your foot and loose sand will spill into the hole to obliterate any interesting features that might have been there. As you approach the beach, the sand becomes both wetter and firmer underfoot. This makes for easy walking, which explains why most people on a beach will walk along the edge of the waves. In this sand, a clear set of footprints will be left behind. If you continue walking seaward into the mushy sand under the water's edge, your footprints will once again become indistinct. Clearly, there is something about the interaction of sand and water which governs the firmness of the beach surface.

This isn't such a strange conclusion. Many materials change character radically when even very small amounts of water are added to them. Salt which flows freely during the winter may suddenly refuse to leave the shaker during the humid summer. Sugar may form clumps in the summer, perhaps even develop a crust because of the moisture in the air. Snow, which may refuse to pack in dry, cold air, even to the point of making winter activities like the building of snowmen impossible, may change character completely when the temperature rises a few degrees. This is because near the freezing point the pressure of your hand is often enough to melt some of the snow, introducing liquid water between the flakes. If my boyhood memories serve me correctly, it seems that the best conditions for making snowballs (including the dreaded "Chicago Iceball") occurred at temperatures of 25 to 30° Fahrenheit.

Footprints in the sand and snowballs are interesting, but understanding them would hardly seem to be a major goal of science. Yet it happens

Footprints in the sand.

they are simple examples of a much wider class of phenomena studied in a field known as soil mechanics. Both sand and snow are examples of bulk materials that are composed of a large conglomeration of small constituents—sand grains in one case, snowflakes in the other. All soils, from the hardest baked clay to the softest mud, share this kind of structure. Thus the lessons we can learn from our footprints will stand us in good stead when we think about building a house, a highway, or a dam. Whether soil will pack and compress under the weight of a structure or exert high pressures on its walls is of more than passing interest to the engineer and architect.

The single characteristic of soils which does more than anything else to determine their behavior is the size and shape of the grains. We know that the names of different kinds of soils (gravel, sand, clay, silt) are given according to the average sizes of the grains of which they are composed. In practice, grain size is usually determined by putting the soil into a sieve and seeing if the grains come through. Such testing takes no account of the fact that the shapes of grains differ. Sand grains, for example, are roughly spherical, while clay is composed of particles that are largely long and flat, roughly the shape of a skateboard. It seems obvious that the way the grains will pack together will depend on the details of their shapes, and it seems reasonable that the properties of the soil should depend on the packing, so the question of whether a foundation will collapse or not often comes down to the question of how the grains in a soil behave.

We can take sand and clay as two familiar types of soils whose behaviors are quite different. One difference, for example, can easily be seen by comparing the behavior of a mud pie to that of a sand castle. When a mud pie dries out, it shrinks and cracks, a fact that tells us that the grains of clay are not very tightly packed together. If they were, the volume of the clay wouldn't change much when the water was removed. A sandcastle may get flaky and not hold together very well when it dries out, but there are no cracks or other signs of shrinking. Sand grains must be tightly packed together.

Clay—even dry clay—has the property of being "plastic." If you take some and squeeze it in your hand, the clay will retain its new shape once you let go. In fact, this "squeeze test" is one of the rough and ready field techniques used by engineers to classify soils. In contrast, dry sand will not retain its shape, but will fall through your fingers as soon as you open your hand. It is not plastic.

We can start thinking about the properties of clay and sand by taking the simplest case—one where there is no water present. We know that if we are to get down to basics, the first question we have to ask ourselves is this: What forces operate on a single grain in a dry soil? Once we know the answer to this question, we ought to be in a position to decide what an individual grain will do, and from this we should be able to deduce what the behavior of the soil will be. We can start with a grain of sand nestled among its fellows.

There are, of course, the usual forces of contact and friction where the grain touches its neighbors (see figure 11-1, p. 144), along with the downward force of gravity. In addition, there are forces that are purely electrical in nature. On the average, of course, the grains in a soil are electrically neutral; they contain as many positive charges as negative. This fact alone, however, does not prevent them from interacting with each other electrically. We saw one such interaction for simple atoms when we talked about induced dipoles in chapter 8. When atoms are incorporated into molecules or locked into solid structures like grains, the possibilities of this type of interaction increase enormously.

When atoms combine to form a solid, you can picture the result as being something akin to a large Tinker-Toy structure. The atoms themselves are represented by the spheres where the sticks come together, while the sticks themselves represent the forces which hold the atoms in place. Depending on the type of material and the number of different types of atoms involved, these structure can get very complicated. The forces between atoms in the structure can be generated in a number of ways. For example, electrons can jump from one atom to another, leaving the original atoms with a net positive charge and the final one with a net negative charge. In this case, simple electrical attraction holds the material together. The atoms in ordinary table salt are bound together this way. Alternatively, atoms may share electrons, a process which creates an attractive force. Most organic materials are held together by this process. Intermediate situations, in which electrons are partly exchanged and partly shared, can also arise.

Thus, we see that individual atoms or groups of atoms in the solid may carry a net charge even though the solid itself is neutral as a whole. To take a somewhat unrealistic example, imagine a crystal made of successive sheets of positive and negative ions as shown on the left in figure 12-1. The total charge of the crystal is zero, but its top side is negative while the bottom side is positive. Two grains of this sort of crystal would experience an attractive force if they were arranged as shown in the center in the figure, a repulsive force if arranged as shown

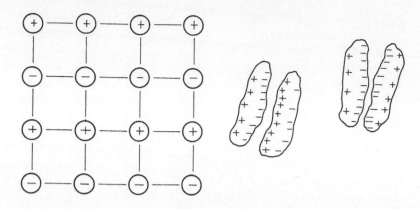

Figure 12-1.

on the right. In general, then, the surface of a grain may have patches of alternating positive and negative charge, and this will give rise to electrical forces between the grains.

It is not necessary that there be patches of charge on grains in order for electrical forces to be exerted between them. For one thing, we saw in chapter 8 that electrically neutral systems such as atoms can exert forces on each other because of the distortion of their shapes on close approach. We saw that the atom, while remaining electrically neutral, could nevertheless have an excess of positive charge on one side and an excess of negative charge on the other. This asymmetry in the charges led to a force between neutral atoms, and, in turn, to phenomena like cohesion, adhesion, and surface tension. Such forces can exist also between atoms near the points of contact between neighboring grains.

A somewhat more common type of electrical forces between grains arises from the configuration of molecules. Like the individual atoms from which they are made, individual molecules are usually electrically neutral. Water is a good example of this. It is made from two atoms of hydrogen and one of oxygen (the familiar H_2O), all of which are electrically neutral and symmetric. When they come together to form a molecule, however, the electrical forces between them are such that the resulting water molecule is highly asymmetrical, as shown in figure 12-2. Furthermore, the electrical forces within the molecule are such that the electrons that were originally orbiting the hydrogen nuclei are pulled over toward the oxygen. (Water happens to be one of those structures where the electrons are partly shared and partly exchanged.) The effect of the electron configuration is to leave the hydrogen end

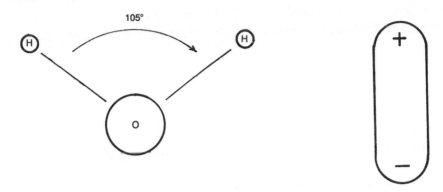

Figure 12-2.

of the molecule with a net positive charge and the oxygen end with a net negative charge. The water molecule, then, has a permanent separation of charges. More complicated molecules, including many that are found in common solids, suffer the separation of charges by the same process as occurs in water. Unlike the separation in atoms, however, the separation of charges in a molecule will be present whether another molecule is nearby or not. Thus, asymmetric molecules can exert electrical forces, and these forces can affect the interactions between grains of source.

The net result is that there are three sort sof forces acting on grains in dry soil: gravity; forces of contact (including friction); and electrical forces. Of these, the first and last determine the important movements of the grain.

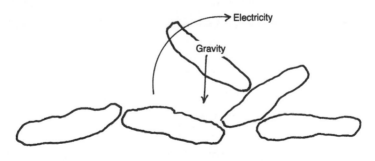

Figure 12-3.

We can look at a sand grain falling into a bed of its fellows to see how these forces operate (see figure 12-3). When the falling grain first contacts one of the stationary grains, all three forces mentioned above act on it. The electrical forces are attractive—indeed, if there were no gravity, the electrical forces would "freeze" the falling grain in the position it had when it first made contact. But the force of gravity is acting, and it exerts a downward force on the grain, tending to make the far end twist around the contact point. The question before us, then, is quite simple: Which force wins? If the electrical forces are stronger than the gravitational, the grain will stay in the position shown. If gravitation wins out, however, the grain will twist and be wedged tightly in amongst the other grains in the bed. As it turns out, the comparatively large size of sand grains and the concomitant large weight ensure that the gravitational forces on the grain overcome the electrical, and the grain falls. Thus, dry sand is effectively compacted by the force of gravity.

This explains why dry sand is not plastic. If you squeeze tightly packed grains like the ones shown above, all you can do is deform the sand grains slightly—they are already about as tightly packed as they can get. Since sand is usually pretty hard (imagine trying to deform quartz), external pressures are unlikely to change the internal arrangements of the grains very much. When you release the pressure, the grains regain their original shape and everything is as it was before.

With clays, the situation is quite different. The same sorts of forces act, of course, but the clay grains are much smaller than their counterparts in sand. Clay particles, by definition, can be no larger than .004 mm—several hundred times smaller than a typical sand grain. If we imagine, for the sake of simplicity, that a grain of clay and a grain of sand are in the shape of a cube, then each side of the clay cube will be at least one hundred times smaller than the side of the cube of sand. This means that the volume of the clay grain will be $(100)^3 = 1,000,000$ times smaller than that of a grain of sand. Since the weight of a given volume of clay and sand aren't very different, this means that a typical clay grain will have only one-millionth the mass of a grain of sand, and that the force of gravity acting on it will be smaller by the same amount.

Thus, when a clay grain falls on a bed and is momentarily halted by electrical forces as shown in figure 12-4, gravity does not win out. The grain simply sticks in the orientation it had when first contact was made, and the result is that the structure of clay has the "house-of-cards" aspect shown.

This explains why even dry clay is plastic. External pressure can push the grains closer together, usually by bending the relatively thin sheets

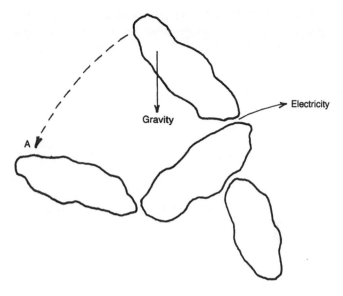

Figure 12-4.

over on themselves. For example, the original grain on the left might be bent over until its free end made contact with another grain at the point labeled A. When the pressure is released, the electrical forces at A will keep the grain bent. It will not spring back, so the clay, once compressed, will keep its shape.

It would be pleasant to be able to report that our understanding of materials had reached the point that, given the number of each kind of individual atoms present we could predict the properties of the grains, and given the properties of the grains we could predict the properties of the soil which is made up from those grains. Unfortunately, our present knowledge is not up to either of these tasks. It's not that we don't understand the basic physics—atomic structure is a mature science. It's just that, as we saw when we talked about the forces on individual sand grains, simple principles are not always easy to apply in the complicated real world. As it turns out, much of our knowledge of the behavior of solids and soils is still largely empirical. This is especially true when we talk about what happens when water is added to the soil, as we must if we want to know why we can leave footprints at the beach.

Sand grains are made of materials that do not absorb water easily. This means that whatever effects water has must be due to its interaction with the impermeable surface of the grain. We know, however, that both the surface of the grain and the water molecules contain displaced electrical charges. It should not be surprising, therefore, to learn that the important effects of water have to do with the electrical forces exerted between the water and the surface of the grains.

Having made this point, I should add that the exact details of this interaction aren't well understood at present. This is partly due to the complexity of the situation, but even more to the fact that many more buildings are put up in clay than in sand, so the former material has received the lion's share of the attention of soil scientists. In any case, the best thinking on this subject now seems to be that the electrical interaction between the grain surface and the dipoles in the water serves to make the water stick to the surface. The water is thus immobilized. Since sand grains start out as a closely packed assortment in any case, when water seeps into previously dry sand there is almost no place it can go where it is out of range of the electrical influence of the surface of some grain or other. In effect, the immobilized water acts as a sort of weak glue, the water molecules serving to apply electrical forces to keep neighboring grains together.

These electrical forces are not strong enough to withstand much external pressure, so when you step on wet sand or try to shape it into a seaside castle, you can make the grains slide over each other so that the sand takes on a new shape. Once the pressure is released, however, the electrical forces hold the grains in their new positions. This explains why Robinson Crusoe was able to see the footprint long after it was made and why sandcastles will stay on a beach until the next tide comes in.

As you walk farther toward the sea, the sand underfoot becomes wetter and wetter. At some point, it becomes so mushy that it can no longer hold the imprint of your foot. What has happened here is that so much water has entered the sand, the grains have been pushed apart. Each grain is still surrounded by a zone of immobilized water, of course, but now there is enough water in the system so that their zones no longer overlap. Consequently, the grains with their immobilized zones are free to move around and the sand ceases to be able to hold its shape.

This picture of the interaction of the sand grains with water explains many of the observations we make at a beach. It explains, for example, why it is easier to build a castle with one sand than with another, since either too little or too much water can make the material hard to work. Similarly, as a sand castle dries out, it does not shrink or crack. All that happens is that water leaves the spaces between grains and evaporates from the surface. The result: the outer part of the castle dries out and the sand either falls off or is blown away by the wind. This process can be countered by periodically sprinkling the sand surface—a procedure regularly followed by "professional" sandcastle builders in Southern California. The picture also explains why the seemingly solid structure falls apart when the first waves hit it, since this results in each grain, complete with its zone of immobilized water, being floated off.

Before leaving this topic, I should make one more point. The possibility of building a sandcastle is sometimes attributed (even by authors who should know better) to surface tension effects. The argument is that if there are two grains of sand close to each other, as shown in figure 12-5, and if there is a little bit of water near the point of contact, then the water will exert a force on the grains as it tries to pull together under the influence of the cohesive interactions that give rise to surface tension. (These forces are discussed in chapter 8.) This statement is true, and such forces do exist in sand that is very slightly wet. However, if we actually perceive the sand to be wet, there is enough water in it to fill the intergrain voids completely. We know that surface tension depends on the fact that at a liquid surface there are attractive forces pulling molecules back into the bulk of the liquid. So surface tension forces of the type pictured could, at best, be exerted only on grains in the outer layers of sand. Only in these layers will there be a surface with water on one side and air on the other. The great majority of grains, however, are in the interior and completely surrounded by water. For them, the primary forces we have to take into account are electrical, and surface tension plays no important role.

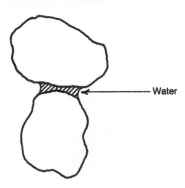

Figure 12-5.

The sandcastle illustrates one of the most important difficulties encountered by engineers who have to erect structures to be supported completely or in part by soils. The physical properties of soils can change drastically when their water content is modified. Sand, for example, becomes plastic when a small amount of water is added. Adding more water does not increase this property, but destroys it completely, as we have seen. The behavior of clay can be even more complex, for chemical reactions between the water and the minerals can become important. Since almost every building is either in contact with or supported by soil of some kind, and since the underground movement of water is difficult (if not impossible) to control, the behavior of water-soil mixtures is of constant concern to anyone concerned with construction.

There is a saying among architectural historians that if a building will stand up for five minutes, it will stand for five hundred years. This is true, *provided* the building has something to stand on. Building materials like stone and steel—even wood—have remarkable powers of endurance if they are properly shielded from the weather. But all the structural strength in the world won't prevent catastrophe if the soil on which the building rests starts to misbehave. Witness the Leaning Tower of Pisa.

One way of visualizing the behavior of a building resting in the soil is shown in figure 12-6. The downward force of gravity on the building

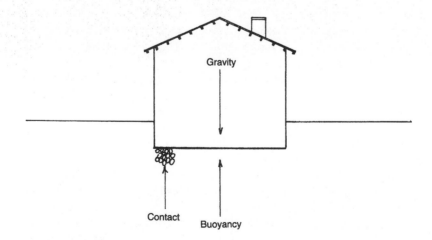

Figure 12-6.

is balanced by two separate upward forces. One, the contact force, is transmitted into the earth through the contact points of the soil grains. The second is the upward force of buoyancy exerted by groundwater. Although it may seem surprising that a building might be subject to the same force that keeps a stick afloat, the pressure force in a fluid can act on concrete as well as on wood. The buoyant force on concrete wouldn't keep it on the fluid surface, of course, but it will make it lighter when it's submerged than it would be when it's on dry land. A building will remain stationary so long as the upward forces of buoyancy and contact cancel the downward forces of gravity.

After a building is put up, however, its weight acts to cause subtle changes in the two supporting forces. We have already talked about the fact that external pressures such as the weight of the building can cause a major rearrangement of grains in clay. They can also cause a somewhat lesser compacting in sand, largely through a small deformation of the sand grains themselves as they are pressed together at the points of contact. In addition, the weight of the building can act to push water out from under it—a process you can easily see operating when you make footprints at the beach. This changes the buoyant force somewhat but, more importantly, the change in water content can have drastic effects on the characteristics of the soil as well. The net effect of the interaction of the building with the soil on which it rests is known as settling.

Every building settles a little after construction; it's an unavoidable consequence of building on a complex structure like soil. Usually, the entire building sinks an inch or two into the ground and no one is the wiser. Occasionally, however, things can go well beyond this sort of modest adjustment. The world record for settling probably was that set by the Museum of Fine Arts in Mexico City, where the entire building sank about twelve feet! Visitors were supposed to walk up a flight of stairs to enter it in the original design; they now enter after walking down a stairway from the street. Fortunately, most of this settling was uniform, so the building is still in use.

In general, the larger the grains in the soil, the faster the water can flow through and the quicker the final stabilization can occur. In some sands, for example, the soil will reach its final state of compaction and water content in a matter of minutes after the load is applied. This probably explains the phenomenal stability of so many of the temples and tombs of ancient Egypt, most of which were simply built on trenches dug in the dry sand. The soil had ample time to respond to each new

Consequences of a poor foundation. The railing was initially straight, but now has a three-inch step. The building is located at a major mid-Atlantic university.

course of stone as it was laid, so that any uneven settling could be compensated during the construction itself. One of the old-time stonemasons in Albermarle County, Virginia (where I live), once told me that when he put up a fireplace he liked to lay a few feet of stone and then let things sit for a month or so before going higher. It took longer to build things this way, of course, but whatever early settling was going to take place could be corrected so that the final chimney stood true. Like the Egyptian temples, his chimneys are still standing and are likely to do so for quite some time.

In clay soils, the situation is somewhat different. There the water moves slowly, and it may take hundreds of years for the soil to respond to a new weight. Settling can occur over that entire time scale, although the greatest amount of sinking will take place immediately after construction (for much the same reason that you get most of the water out of a sponge on the first squeeze). A famous and well-studied example is the great entrance rotunda at Massachusetts Institute of Technology, in Cambridge, Massachusetts, a structure known as "Building 10."* Finished in 1910, Building 10 immediately showed an alarming tendency to settle, with various portions sinking a full five inches in the first decade of its existence. Aside from the embarrassment of having the main entrance to a prestigious school of engineering sink into the Charles River mud, the fact that one side of the portico had settled an inch more than the other suggested some rather alarming consequences if the trend were to continue. Fortunately, by the time the future looked its bleakest, most of the adjustments in grain contacts and water levels had been made, and the settling has been quite modest since 1925, so that future generations of students and visitors can go into the building without fear.

---

* I spent a very enjoyable and productive year as a postdoctoral fellow at MIT and I yield to none in my respect and admiration for that institution. I did wonder, though, whether giving the buildings names instead of numbers would have detracted from MIT's macho, high-tech image.

13

SAILING SHIPS AND SKELETONS

Our sails they were good and strong, made of the finest duck
Our ropes were of manila and passed through painted block
Our vessel made of white oak, and finished with great taste
To ride the heavy Norther gale and stand the winter's test.

—"The Ocean Queen,"
New England Whaling Song

Near any population center on a fine weekend, the ocean seems to sprout a forest of white as sailboats come out of the harbors and marinas. There is a great deal of pleasure in sailing, and some nostalgia as well. Until a comparatively short time ago, most of the world's long-range commerce was carried out by sailing ships. The triumph of steam and the disappearance of the clippers (perhaps the ultimate sailing ships) is less than a century old. I suspect that a good portion of "down to the sea in ships" nostalgia is misplaced, for life on a sailing ship must have been primitive at best. Nevertheless, there is something in the human spirit that derives enjoyment from contemplating idealized versions of a lifestyle which is now safely and irretrievably in the past, so that sports like sailing and horseback riding will undoubtedly be popular for a long time to come. There is a pedagogic value to sailboats as well, because they are living, moving examples of Newton's Laws of Motion and the vector nature of force.

The earliest sailboats were primitive affairs, hardly more than rowboats with a sail that could be hoisted to catch a favorable breeze. Such boats are simple to understand: the wind pushes on the sail, moving the boat downwind. Anyone who has ever been caught in a gusty storm with an unfortunately positioned umbrella is familiar with this effect. In sailing parlance, a boat that is being pushed along in the direction of the wind is said to be "running before the wind."

It often happens, however, that you want to have the boat move in a direction different from that of the wind. Even the earliest sailing boats in recorded history, boats depicted by Egyptian rock carvers in 3300 B.C., had to deal with this problem. Their solution, which involved a large steering oar, made use of the modern concept of the vector, although the Egyptian helmsman would surely not have thought in those terms.

The term "vector," as used in physics, refers to quantities which have both a size (or magnitude) and a direction. The most familiar vector is probably velocity. If you want to specify the motion of a car precisely, you have to say something like "It's going north at fifty miles per hour." Both pieces of information, the magnitude (fifty m.p.h.) and the direction (north), have to be given if we are to have an accurate notion of the velocity. Either by itself is incomplete, as you can readily see by thinking about the statements, "The car is moving north," and, "The car is going fifty miles per hour." Neither of these statements describes the motion fully. (I should note that this use of the term "velocity" in physics is somewhat different from the ordinary colloquial usage. For the magnitude of the velocity vector—fifty m.p.h. in our example— physicists customarily use the word "speed," reserving the term "velocity" for magnitude plus direction.)

Force is another vector quantity, having both a magnitude and direction. If I say that I'm pushing with a force of fifty pounds on a chair, you can't tell which way the chair will move until I tell you whether the force is directed upward, horizontally, or down. In the case of a sailboat running before the wind, the force of the wind in the sails is clearly in the same direction as the wind itself, which is why that case seems easy to analyze.

Vectors have the property that they can be added and decomposed.

To understand what this means, picture two individuals trying to move a heavy load by pulling on two ends of rope, as shown on the left in figure 13-1. Clearly, each is exerting a force on his rope, and the direction of that force is along the rope. Thus, the object feels two forces being exerted, one in the direction of each of the people doing the pulling.

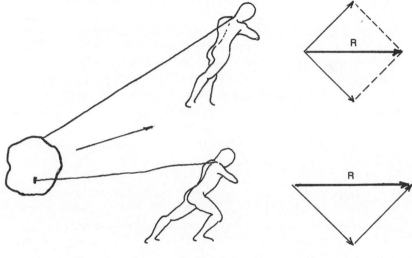

Figure 13-1.

Common experience tells us that the object will not move toward either person, but in the direction indicated by the arrow in the figure. We know from Newton's second law (see p. 143) that the rate of change of velocity (acceleration) of any object is in the direction of the force being applied to it. From this, we conclude that the effect of two forces acting in different directions on an object is to produce a force acting in yet a third direction.

One way of thinking about this fact is shown in the center of the figure. If we represent the two forces being exerted by the ropes as arrows, with the length of the arrows representing the magnitude of each force, then the actual force felt by the object is represented by the arrow labeled R. If you imagine the two arrows representing the applied forces as being two sides of a parallelogram, and if you complete the parallelogram with the dotted lines as shown, then the vector R is just the diagonal of the resulting figure. (This construction is called the parallelogram of forces.) Another geometrical mode of thought is to imagine sliding the upper arrow along the lower until they are head-to-tail, as shown on the right in the figure. In this case, R is the third side of the triangle, whose other two are the applied forces.

These geometrical rules can be used to add up (or, in technical terms, compose) any two vectors. The fact that the motion resulting from the application of two forces is not necessarily in the direction of either force explains many common phenomena. For example, it is often necessary, when walking along the side of a steep hill, to aim your steps in a slightly uphill direction in order to keep at the same level. This is because the force of gravity adds a slight downhill vector to the horizontal force applied by your foot to the ground. If you didn't allow for this extra vector, your motion would also be slightly downhill.

The fact that vectors can be added has a corollary that may seem a little strange at first glance. If it is true that any two vectors can be composed to give a single result, it is equally true that any single vector

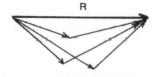

Figure 13-2.

can be considered the result of the addition of two other vectors. In figure 13-2 we show how a single vector can be thought of as the sum of any number of pairs. If you think of each pair as a set of ropes being pulled with varying force, this idea should seem reasonable. Just as two forces in figure 13-1 can be replaced by the single force we labeled R as far as the motion of the weight is concerned, any single force that acts can be thought of as the result of two other forces. When we think in this way, we say that we decompose the original vector, and the two forces are called components of the original vector.

With this background, we can rejoin our Egyptian helmsman on his boat. Running before the wind is an easy situation to analyze in terms of vectors. When the wind strikes the sail, it imparts a forward force to the ship. As the boat moves through the water, there is a frictional force which tends to oppose the forward motion. Both forces act along a line in the direction of motion of the boat, so the addition of the vectors is quite simple. If the wind has just arisen and the boat is starting up, the forward force will be larger than the frictional one, and the result will be as shown on the left in figure 13-3. There will be a net forward force, and the boat will be accelerated in the direction of the wind.

Figure 13-3.

After a while, we will have the situation shown in the center. The forward force due to the wind will be exactly canceled by friction, so the net force on the boat will be zero. In this case, it will just move along at constant velocity, since there is nothing to make it speed up or slow down. Finally, if the wind should suddenly die, we would have a situation like the one shown on the right, where the frictional force would dominate. In this case the boat would slow down and eventually stop.

If the Egyptian helmsman wanted to change direction, he fitted the ship with a large steering oar in the stern. By putting this oar into the

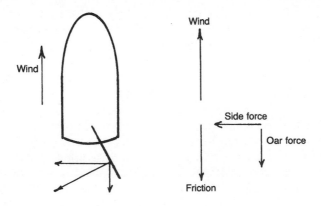

Figure 13-4.

water at an angle, it was possible to generate a sideways force on the boat. The ultimate origin of this force is the collision of water molecules with the oar, and the force will always be perpendicular to the oar. We can, however, decompose this force into two components, as shown on the left in figure 13-4, one in the direction of the wind and one perpendicular to it. At the moment that the oar is dropped into the water, the four forces on the boat are as shown on the right. One component of the force on the oar acts in the same direction as friction and tends to slow the boat down. The other component is perpendicular to the direction of motion and will cause the boat to accelerate sideways. This acceleration will continue until the friction associated with the sideways motion cancels the force exerted by the oar.

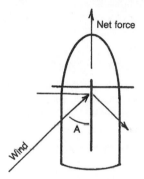

Figure 13-5.

Although we've only discussed the case where the wind is coming over the stern, a simple, fixed sailboat like this can make headway if the wind is coming at an angle as well. As shown in figure 13-5, in this case the wind can produce a net forward force by bouncing off of the sail; but the force of the wind on the side of the boat will produce a force tending to move the boat to the side. The larger the angle between the sail and the wind, the worse this becomes, and the harder it is to compensate with the steering oar. While it is possible in principle to extract energy from the wind up to the point that the angle labeled A in the figure reaches 90° ("wind abeam"), as a practical matter the wind will cease to be of use at much smaller angles. To deal with such an emergency, boats from Egyptian times until midway into the middle Ages came equipped with oars. If all else failed, old-fashioned elbow grease would pull them through.

It's clear from this discussion that an Egyptian ship could never sail "into the wind." The ability to do so—an ability that had to be acquired before serious ocean commerce could begin—was developed in Europe by the Vikings. A ship has to have two characteristics to be able to move into the wind: it must have a movable sail and it must have a keel. We shall see why shortly. The movable sail can be of the type seen on modern pleasure craft, where a triangular sail is attached to a boom which can swing back and forth; or it can be achieved by adjusting the lines that hold the ends of square sails suspended from a bar perpendicular to the mast. In the latter case (which was used by the Vikings and clippers alike), the sails are obviously less mobile, but this is compensated for by the fact that they can be much larger and thus produce a greater forward impetus.

Lest there be a misunderstanding, I should point out that no sailing boat, no matter how well designed, can actually sail into the eye of the wind. What is possible, however, is to guide the boat at an angle of less than 45° into the wind. As shown on the left in figure 13-6, there is a "dead area" extending roughly this far on either side of the wind direction. This area cannot be entered (although some exceptionally well-designed and skillfully handled boats can get as close as

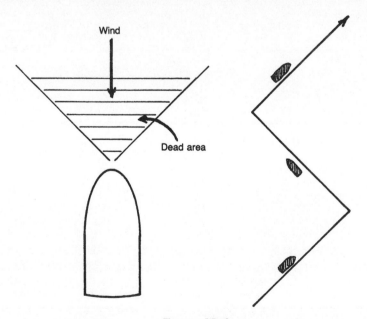

Figure 13-6.

35°). Sailing into the wind involves a series of right-angle turns such as those shown on the right, a maneuver known as tacking.

But this talk of maneuvering presupposes a solution to the question: How can the wind blowing on a sail produce a force which moves the boat against that same wind at any angle? How, in other words, do the sail and keel overcome the Egyptian helmsman's limitation?

We can start by looking at the sails. Sails are never stretched taut, but are shaped so that they billow out. This is not an oversight but a clever way of getting more useful energy from the wind. When the wind flows around a sail (see figure 13-7), two forces are generated. One is the standard kind of push that we have talked about up to this point. It acts perpendicular to the surface of the sail at each point. The

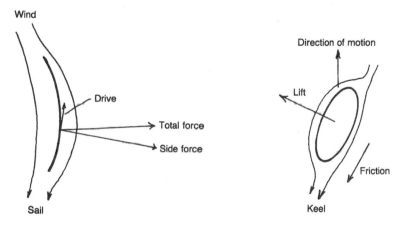

Figure 13-7.

second force is familiar to us from the discussion of the forces on a sand grain in chapter 11. Since wind moving around the outer part of the sail must travel a longer distance than the wind which moves along the inside, there will also be a lift force present, exactly the same lift force that keeps an airplane in the air (see p. 147). The net result of these two effects is that there is a force on the sail in the direction shown. For convenience, we break this force into components: one called "drive" in the direction of the long axis of the boat, and the other a side force perpendicular to the drive.

If these were the only forces present, the ship would slide sideways through the water at a small angle to the wind. However, the sail is only half the story in a sailing ship. The underwater keel also produces

was upright, these extra wings had very little effect. Their usefulness became apparent in the tacking maneuvers. As the boat heeled over because of the action of the side force and lift just described, the wings made it more difficult for the boat to slide sideways in the water. In effect, they increased the vessel's resistance to motion in a direction perpendicular to its main axis. Thus, the *Australia II* was able to sail closer to the wind with a smaller keel than could a ship of conventional design. This wasn't much of an advantage, but it was enough to move the America's Cup to Perth.

All the forces we have discussed so far involve external agencies acting on the sailboat. There is another set of forces that are equally interesting—those that act within the boat itself. The central question in building a sailing vessel is to decide how to support the mast and the weight of the sails. Two answers present themselves, sketched in figure 13-10. One, widely used in smaller boats, is to make the mast heavy enough, and to fasten it to the hull strongly enough, so that it is capable of holding the sails without further support. The second answer, universally employed in large ships, makes use of ropes and cables to support the mast. As it happens, general principles of nature explain why each solution fits its particular case.

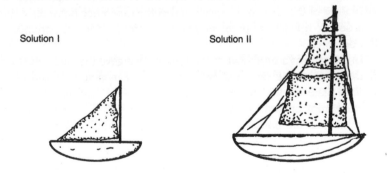

Solution I   Solution II

Figure 13-10.

A structural member like a mast or a column, which carries a load in such a way that the force acts to push the atoms together, is said to be in compression and is usually referred to as a "compression member." Ropes and cables, where the forces act to pull the atoms apart, are said to be in tension and are called "tension members." The strength of a material or structure under tension and under compression is not necessarily the same. A rope, for example, has no strength under compression at all, but can hold weight under tension. Similarly, a brick wall with mortar joints can hold a very heavy load under compression, but the mortar will quickly give way under tension. The job of designing a ship (or any other structure) thus comes down to choosing the right materials and the right mix of compression and tension members to do the job in the most efficient way. In practice, being "efficient" means doing the job with the least amount of material and expense. In the design of racing boats, reducing the amount of material (and therefore the weight) is probably the most important single criterion in design.

The weight of a member needed to carry a specific load is quite different for compression and tension. If we hang a ten-pound weight from a rope, there is a ten-pound force of tension everywhere along the rope. It makes no difference if the rope is one foot or fifty feet long, the tension is the same. This means that if we want to go from a small ship to a large ship, all we have to do is increase the length of the ropes; we do not have to make the ropes thicker. This will increase the weight of the ship, of course, but only in proportion to the increase in size. Doubling the length of a rope to accommodate a larger design just doubles the rope's weight.

The situation with compression members is different. If we want to increase the height of a mast or column, we have to worry about the fact that the column may buckle when its height reaches a certain value, even if no new weight is added to it. You can convince yourself of this fact by doing a simple experiment. Take a few drinking straws. A single straw will easily support the weight of something like an orange, but a column made by joining several straws together will quickly collapse. The only way to keep a taller mast from buckling is to make it thicker as well as taller. For example, if we want to double the height of a cylindrical mast while preserving its structural properties, we also have to double the radius of the cylinder. This means that if we want to make a mast twice as tall, we have to make it eight times as heavy. In contrast with tension members, there is a severe weight penalty for increasing the size of compression members. This explains why the designers of tall sailing ships invariably choose the second solution to the problem of supporting the mast. Adding extra ropes and cables is cheap, while strengthening the masts themselves is costly.

Actually, the ship designer isn't the only one who has taken note of this simple law of physics and chosen to build structures with a few compression members held in place by a collection of members in tension. Mother Nature, in designing the human body, has done the same thing. The compression members of the body, the parts that carry the weight, are the bones that make up our skeletons. As with the tall ships, our compression members are few in number and supported by a large number of "ropes and cables" in the form of muscles, tendons, and ligaments. There is more behind the design of a sailboat than meets the eye!

Another example of a structure where we can see the physical principles of the sailboat in action is the medieval cathedral. These buildings were the first to use a skeleton-like construction in which the main weight of the roof was supported by large columns, leaving the space between the piers to be filled with whatever material the architect chose to use. Since there was no load to be carried by these materials, even glass could be used—which explains the superb stained-glass windows that contribute so much to the sense of awe one feels in these buildings. To help support the weight on the columns, a "flying buttress" (see figure 13-11) was often used. In such a structure, the downward force of the roof must be balanced by the upward force of the pier and the buttress. Newton's second law tells us that if the buttress and pier exert a force on the roof, the roof exerts an equal and opposite force on the

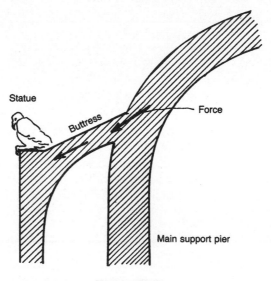

Figure 13-11.

buttress and pier. This reaction force is indicated by the arrows in the figure. The flying buttress carries this force to the left until it comes to its own support pier. The support is supposed to carry the force to the foundation in safety. But a question occurs: Why do so many flying buttresses have large ornamental statues on top of their support columns?

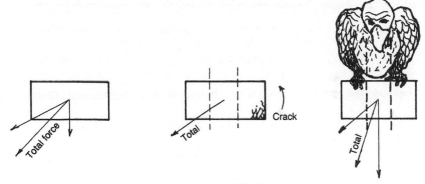

Figure 13-12.

The reason might be aesthetic, but are there practical reasons as well? It turns out that there are. Brick and stone are very strong under compression. They do not, however, easily support a load imposed at an angle. To see why, consider the top block of stone in the support pier of the flying buttress figure 13-12. The forces on this stone are shown on the left; they are, respectively, the oblique thrust of the roof and the downward weight of the stone itself. The net force on the block is obtained by the usual technique of vector addition, as shown in the figure.

The important point about masonry structures is this: If the total force vector falls outside the middle third of the block, as shown in the center of the figure, the block will start to tilt over, placing the mortar on the inner edge of the pillar under tension. This causes the mortar at that point to crack. If the vector falls outside the middle third of the block, then, the column will start to bend. You often see this effect in the chimneys of old houses.

One way of preventing the cracking of the joints is to put a heavy statue on top of the block. The statue's weight is then added to the forces of the block, as shown on the right in the figure. If the statue weighs enough, this extra downward force will be enough to pull the total force vector back inside the middle third of the block. Paradoxically, the column is made more stable by adding weight to the load it must carry.

Medieval builders obviously did not know the laws of vector addition, but they had a long tradition of experience in putting up buildings that would stay up. If it were not for their grasp of the facts that we call vector addition, they would not have been able to hold up the roofs of their cathedrals with walls so thin that they could be pierced by huge windows to let in the light, and we would have been deprived of the beauty of the stained glass of Chartres and Notre Dame.

14

SKIPPING STONES

To myself I seem to have been only a little boy playing on the seashore, and diverting myself in now and then finding a smoother pebble or a smoother shell than ordinary, whilst the great ocean of truth lay all undiscovered before me.

—Sir Isaac Newton

Newton had no children. If he had had, he might not have treated so cavalierly the activity of finding stones and pebbles on a beach. There are few pleasures in life greater than passing on to your children the knowledge of how to pick rocks which have just the right shape to skip along the water when thrown. In this age when a parent's experience is so often irrelevant to his children's needs, it's comforting that a simple skill can be passed from one generation to the next. The art of skipping stones is probably one of the few left that is transmitted completely by oral tradition. It almost has to be, because by one of those delightful quirks that occur all too seldom in our scientific age, we do not really understand very much about how the skipping stone works.

This rather surprising fact was brought to the attention of the scientific world in 1957 by a retired professor of English at Columbia University, Ernest Hunter Wright. Writing in the "Amateur Scientist" section of *Scientific American,* Mr. Wright described how he had been walking along a beach and, instead of trying to skip stones on the rough surf, contented himself with skipping them along the wet sand at the water's edge. He noticed that the stones left a characteristic series of marks in the sand. The stone seemed to strike the sand, take a short hop of about four inches, and then take the charactistic long hop we associate with skipping. Wright asked his physicist friends to explain this phenomenon. They couldn't. In Wright's words, "With the luck of a layman, I have had the novel experience of seeing several of the men who plucked the heart out of the atom's mystery scratch their heads in vain for the solution of a problem which I now submit to a wider audience."

This puzzle caught the attention of the scientific community, and more than ten thousand people wrote to the journal offering comments and theories. An important piece of the problem was illuminated by Kirston Koths, then an undergraduate student at Amherst College. He set up a high-speed camera and actually photographed stones skipping on sand and water. His results were interesting, if somewhat unexpected. When a stone skips on sand, it behaves exactly as Wright had said: short and long hops alternate. On the contrary, when a stone skips on water, there are only long hops. It's clear from the photographs (which appeared in the August 1968 issue of *Scientific American*) that the closely spaced marks in sand are made by the back and front end of the stone, respectively, and that the short hop corresponds to the stone being tipped over by the hard sand surface. In water, the back edge of the stone seems to hit and skim along the surface, building up a mound of water in front of itself before taking off on another hop. This corresponds to what you see when you skip a stone—it travels in a series of ever shorter hops, with no evidence of the kind of double contact that Wright talked about. With Koths's work, then, we have a reliable picture of what really happens when a stone skips. The question of why it happens is another matter.

When I was a student, I found the physics of spinning objects to be one of the most difficult subjects I had to learn. Nothing in my teaching experience has given me any reason to think that I am unique in this respect. For some reason, the human mind has a hard time getting hold of things like rotation and spin. Since the most crucial aspect of the skipping stone is the fact that it is thrown so that it spins on its axis, we're going to have to take the plunge and learn a little about the dynamics of rotating bodies.

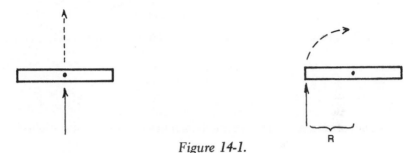

Figure 14-1.

There are two important concepts that we need here: angular momentum and torque. It's obvious that if we apply a push to a stick at its center, as shown on the left in figure 14-1, the stick will simply move forward in response to the force. If, on the other hand, we push off center, as shown on the right, the stick will rotate as it moves forward. A force applied in such a way as to cause rotation is said to exert a torque. The magnitude of the torque in the figure is simply the force multiplied by the distance between the center of the stick and the point of application, the distance labeled r. If the force were applied at an angle instead of as shown, then we should insert in this definition the component of the force perpendicular to the stick.

So far, so good. The difficulty comes not so much from defining the magnitude of the torque as its direction. I won't try to explain why things are done this way, other than to point out that the following procedure is a natural one in a field of mathematics known as vector calculus. The rule is this: If you put the index finger of your right hand along a line from the center of the stick to the point of application of the force, and the middle finger of your right hand in the direction of the force itself, then your right thumb will be in the direction of the torque vector. (To make this rule work, you must hold your thumb perpendicular to the two fingers.) The torque on the right in the figure above, then, points into the page.

The force shown on the left in figure 14-1, because it acts through the center of the stock, has r = 0 and does not produce a torque. It produces no rotation, either, which leads us to suspect that there is probably a connection between the size and direction of the torque being applied and the subsequent rotation of the object. Angular momentum is simply a convenient quantity for describing rotation. Like torque, angular momentum is a vector. If we think of an idealized skipping stone as a rotating disk (see figure 14-2), then the direction of the angular momentum vector is given by another right-hand rule: Wrap the fingers of your right hand in the direction of the spin and your thumb is in the direction of the angular momentum. In the figure, the angular momentum is perpendicular to the disk and points upward as shown. The magnitude of the vector depends on the radius and mass

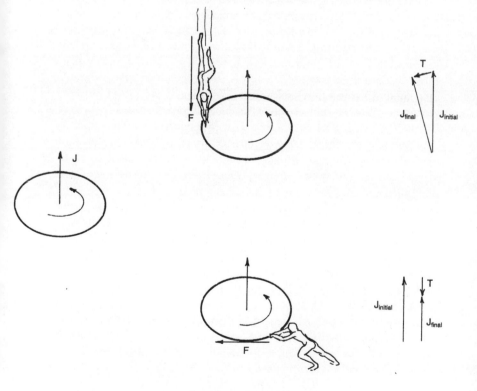

Figure 14-2.

of the disk as well as on how fast it is spinning.* The angular momentum of an object is customarily denoted by the letter J.

With these definitions out of the way, we can turn to the question of the effect of forces on rotation. We know that for linear motion (see p. 143) Newton's second law relates the applied force to the acceleration. Newton's second law for rotation takes much the same form. It says that the rate of change in angular momentum is equal to the applied torque.

We can get some idea of how this torque-momentum relation works by looking at the central diagram in figure 14-2. The force is applied tangentially to the spinning disk, so that the resulting torque is perpendicular to the disk and directed downward. This means that after the force has been applied for a period of time, the change in angular momentum will be represented by the vector labeled T. The addition of the change in angular momentum to the original angular momentum vector is shown next to the disk in the figure. The result—that the final angular momentum vector points in the same direction as the original but is smaller in magnitude—is easy to understand. It simply says that when we apply a force like the one shown, we slow down the rotation. This, after all, is the principle of the automobile brake.

A particularly interesting case is the one in which the force is applied perpendicular to the disk, as shown on the right of the figure. The torque associated with this force will be perpendicular to the original angular momentum. In this case, the final angular momentum must be obtained by the kind of addition of vectors we described in chapter 13. The final angular momentum will point in a different direction from the original, which means that the disk will tilt over in response to the force. If the force is applied for a short time, so that the tilting of the disk doesn't change the orientation very much, then the vector representing the change in angular momentum will be small and there will be very little change in the magnitude of the total angular momentum. In this case, the disk will tilt slightly but will keep spinning at the same rate.

The most important point about the spinning stone, however, is that in the absence of torques, the angular momentum cannot change. This means that as the stone moves through the air, the direction of the

*For a disk of mass M and radius R turning once every T seconds, the magnitude is $\pi MR^2/T$.

angular momentum vector must remain fixed in space, as shown in figure 14-3. We put a spin on the stone, then, to ensure that its orientation does not change between the time it leaves our hand and the time it hits the water (or sand) before the first hop. This particular property of angular momentum also explains why a stone skips best if it is thrown with a side-arm delivery and leaves the hand with the skipping side almost parallel to the ground. If the stone is released in this position, then the angular momentum associated with the spin will ensure that it will be in roughly the same position when it hits the water, and it will not flop or tumble in flight.

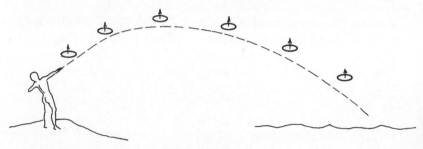

Figure 14-3.

The first place that a torque can be applied during the flight of the stone, then, is when it hits the water or sand. Let's consider sand first. When the stone hits (as shown on the left in figure 14-4, we can think of the wet sand as exerting two forces on it. One of these will be a frictional force caused by the edge of the spinning stone moving against the sand. This force will not be very large, since the sand is wet, but in any case we know that its effect will be to slow the spin and decrease the magnitude of the angular momentum.

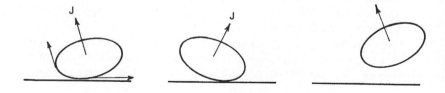

Figure 14-4.

The second force involved is the ordinary force of impact, and is directed perpendicular to the stone as shown. The effect of this force, we know, will be to change the direction of the angular momentum vector. We expect, therefore, that the net effect of these two forces will be to cause the stone to flip over, as shown in the figure. This flipping behavior accounts for the two closely spaced marks that Wright noticed.

At the second impact, the same two sorts of forces operate, so when the stone takes off, the angular momentum vector has been rotated a second time.

When the stone hits water, the behavior is rather different. There is still a frictional force, of course, although it will be even smaller in water than it was in wet sand. But because the water is not solid, the impact of the stone causes a wave to build up under it and the stone skims across the water surface as shown in figure 14-5. Since whatever upward force the water exerts is spread out across the entire surface of the stone, there is a very little change in the angular momentum vector. In fact, from Koths's photographs, it appears that the upward force on the leading edge (labeled F_L in the figure) actually wins out over the force exerted on the trailing edge (labeled F_T in the figure), so that the angular momentum vector is rotated slightly backward by the impact. Why the water behaves in this way is not understood (at least, I haven't seen any explanation of it).

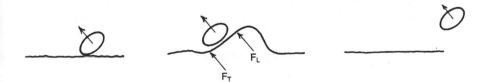

Figure 14-5.

In subsequent hops the patterns described above are repeated, except that on each hop the stone is spinning a little more slowly than it was before and a little more of its energy has been expended in the sand or water. The hops get shorter and lower, until eventually the stone hits in a way that causes it to dig into the surface and stop. My own record is eleven hops on water, made one day when Lake Michigan was particularly smooth. I haven't been able to beat that yet, but with two daughters just entering the stone-skipping age, I'll certainly have a chance to try.

The same principles governing angular momentum in the skipping stone also govern the behavior of many other interesting (and sometimes useful) systems. The gyroscope, for example, is essentially nothing more than a rapidly rotating disk of the type we have been discussing. Once it has been set into motion, we know that the direction of the angular momentum of the disk can be changed only by the action of an external torque. If the gyroscope is mounted in a set of low friction bearings and enclosed in a casing, all the torques can be made very small. Consequently, a spinning gyroscope is a device which can be used to define a fixed direction in space, regardless of the motion of the platform on which it is sitting.

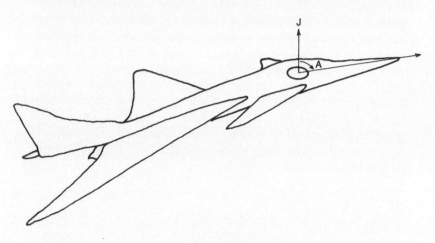

Figure 14-6.

There are many situations in which such a device is very useful. In an airplane or satellite, for example, it is not always possible to determine the orientation of the vehicle by looking outside. An internal gyroscope, on the other hand, can be used to give this information, as shown in figure 14-6. The operator can look at the angle between the vehicle and the direction defined by the gyroscope and deduce the true orientation. For example, if he finds the angle labeled A to be 10°, he knows that the axis of his craft is 10° below the direction defined by the gyroscope axis. He can then act on that information. Sophisticated versions of this scheme, in which the readings are taken electronically and fed into a computer, are at the heart of inertial guidance systems used in aerospace technology.

A somewhat less complicated device for the study of angular momentum is the child's top. A top has several stable modes of operation. It can stand upright and spin, as shown on the left in figure 14-7. The only forces acting on the top (ignoring friction) are those associated with the point of contact and with gravity. In the upright position, neither of these forces can produce a torque about the contact point. This means that the angular momentum of the top, as given by the right-hand rule and shown in the figure, will remain unchanged. The top will continue to spin in this configuration unless acted on by an outside agency.

If, however, the top is started spinning at a slight angle, as shown in figure 14-7, the situation is different. Now the force of gravity on the top is no longer directed along its axis, but at an angle. This means that there is now a torque being exerted on the top, and a few moments of communing with your right hand should convince you that the direction of this torque is out of the page (i.e., pointing toward you as

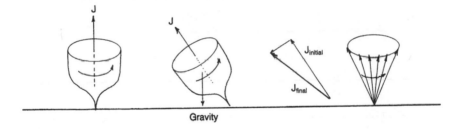

Figure 14-7.

you look at the drawing). This is one of those situations where the torque is perpendicular to the angular momentum, so we expect the effect will be to change the direction of the angular momentum without changing its magnitude. As this affects the top, the equivalent statement is that the axis of the top will rotate but the rate of spin will not change.

As soon as the top has rotated a small amount, however, the direction of the torque will change because the mass of the top has moved. The new torque will be perpendicular to the new angular momentum, and the entire process discussed above will repeat itself. The top will rotate, the direction of the torque will change, which causes the top to rotate again, and so on. The new result is that the direction of the angular momentum will move around as shown on the right in figure 14-7, and the axis of the top will describe a cone in space. This sort of motion is known as the *precession* of the axis of the top. It occurs in any object which is spinning and subjected to a torque that remains perpendicular to the angular momentum.

Anyone who has much familiarity with tops will realize that spin around an axis and precession of that axis are not the only kinds of motion that a top can undergo. Once we get past precession, though, the discussion gets highly technical. The guiding principle relating angular momentum to torque remains the same, but the application becomes more difficult. So I will just describe the motions briefly, without explaining in detail how they arise.

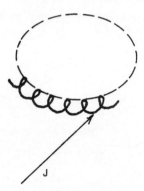

Figure 14-8.

There is another possible response of the top to the torque exerted by the weight, and that is for the axis to precess and wobble at the same time. In this case, as shown in figure 14-8, the tip of the axis describes a looping path centered on the normal motion associated with precession. This extra wobble is termed *nutation*. A top can precess with or without nutation, the choice of the motion depending on the proportions of the top and the rate of spin.

Another motion one sometimes sees is "sleeping" or "rising," in which the axis of the precessing top actually moves up to a vertical position. This effect occurs because the point of contact between the top and the ground, which we have been treating as infinitely sharp, is actually rounded. When the top tilts over to precess, the forces of friction and contact are exerted slightly off center from the top's axis of symmetry. It is the extra torques due to this small asymmetry which cause the top to rise.

Other than being a child's toy and a very difficult physics problem, the spinning top also serves as a model of rotating systems in nature. For example, the elementary particles which make up the subatomic structure of matter can be thought of as tiny tops spinning about an axis. In particular, the nucleus of any atom can be pictured in this way. It turns out that because the nucleus is composed of particles which carry electrical charge, the fact that it is spinning means that it behaves like a tiny magnet as well. As shown in figure 14-9, the nuclear magnet is lined up in such a way that it points in the same direction as the angular momentum associated with the nuclear spin.

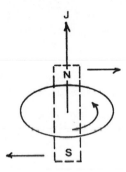

Figure 14-9.

If the nucleus is placed in a magnetic field (for example, by putting it between the north and south poles of a strong magnet), there will be forces exerted on the nuclear magnet. The north pole of the nuclear magnet will be pulled one way, the south pole the other, as shown. By now, you should be comfortable enough with the ideas we have been discussing to recognize that there is a torque acting in this situation, a torque which is perpendicular to the direction of the nuclear spin, the nuclear angular momentum, and the nuclear magnet. Although the origin of the torque is different from what it was in the case of the spinning top, the end result is the same. The spin of the nucleus will precess around the magnetic field. The rate of precession will depend on the torque, and hence on the strength of the nuclear magnet and the strength of the applied magnetic field.

If we have a large number of nuclei in a sample and subject the sample to an external magnetic field, it is possible to use some fairly simple electronics to measure the rate of precession of the nuclear magnets. This technique is known as magnetic resonance imaging (MRI), and its discovery won Felix Bloch of Stanford and Henry Purcell of Harvard the Nobel Prize in 1952. The technique has a multitude of uses, one of the most exciting and most promising being in the field of medical diagnostics.

We know that we can measure the rate of precession of nuclei in a magnetic field, and that each different kind of nucleus has a different rate of precession. Carbon, for example, will produce a different readout in an MRI apparatus than will phosphorus. Furthermore, if the external magnetic field in the sample is not uniform but varies slightly from one spot to another, the readout will not only tell us what sort of atom is involved but also the size of the external magnetic field at the atom's location. This, in turn, allows us to deduce the location of the atoms from the (presumably) known values of the external field.

MRI as a diagnostic tool works in this way: the patient is placed in a large cylindrical cavity where a well-measured magnetic field is maintained. The precession rates of the atoms in the patient's body are then obtained and fed into a computer, which produces three-dimensional images of the patient's internal organs from the information. The pictures produced are similar to those produced by CAT X-rays but, unlike the X-ray images, it is not necessary to expose the patient to risk to obtain them. As this is written, industry spokesmen are predicting that MRI units will be installed in hospitals in the United States at the rate of one thousand a year. It may well be that MRI scans will be the average person's first direct contact with the effects of nuclear spin.

Unlike MRI, which has clear practical and economic value, another example of precession in science takes place in one of the most abstract fields of basic research, general relativity. The original theory of general relativity was first proposed by Albert Einstein in 1915. In 1919, Arthur (later Sir Arthur) Eddington verified one of the predictions of the theory, that the path of light from distant stars would be bent as it passed near the sun. It is a major anomaly in the history of science that the theory was then accepted almost immediately, without having gone through the rigorous experimental testing usually demanded of new ideas. The reason for this, I suspect, is that the theory is so beautiful and elegant that a scientist's first reaction upon seeing it is that it just *has* to be right.

Yet physicists tend to be a little embarrassed by the weak experimental underpinnings of general relatively, so whenever someone can find a new way to test the theory, he generates a lot of interest. The late

Leonard Schiff of Stanford University, working in the early 1960s, found that general relativity made a rather remarkable prediction about the behavior of a rotating object placed in orbit around the earth. According to the theory, there should be a small torque on such an object—a torque which is not predicted by any other theory. This torque has the property that it is perpendicular to the angular momentum of the spinning object, a fact which we now recognize implies a precession of the axis of rotation.

This precession wouldn't be very noticeable—for a spinning sphere the size of a Ping-Pong ball it would take 180,000 years for the axis to make one full revolution. Detecting such a small motion by remote control is no mean feat, to say the least. Nevertheless, Francis Everitt of Stanford began work on designing an experiment to do so in the late 1960s, and in April, 2004, under the name of 'Gravity Probe B', the apparatus was launched.

The experiment will actually consist of sending out four quartz spheres cooled to within a few degrees of absolute zero ($-456°F$). Once the system is in orbit, the spheres will be "spun up" by small jets of helium blown across their surfaces and then will be allowed to coast freely in a near-perfect vacuum for a year, during which time extremely precise measurements of the precession (or lack of it) will be made. So sensitive is the experiment that even very small inhomogeneities in the quartz spheres could easily mask the effect the experimenters are searching for. Consequently, the Stanford group had to find ways of producing what Everitt loves to call the world's roundest object—a sphere the size of a Ping-Pong ball so uniform that if it were blown up to the size of the earth, the tallest mountain would be only a foot high. If they succeed in finding the predicted precession, an important new experimental underpinning will have been added to the theory. If not, the race will be on to replace general relativity with something new.

Our thoughts on the skipping stone have led us far afield. It is clear that physicists know a great deal about angular momentum and precession, so the fact that we still do not completely understand the stone's behavior is probably due to the problem not being regarded as terribly important. Someday, I suppose, someone with time on his hands will take it up and solve it. All I can say is that it won't be me. I prefer the skipping stone to remain a modest mystery in science, at least until I've had a chance to pass on my skill and experience in the art to my daughters.

INDEX

James Trefil, with his daughters
Dominique (left) and Flora (right)

JAMES TREFIL is Clarence J. Robinson Professor of Physics at George Mason University. He is a fellow of the American Physical Society and the American Association for the Advancement of Science, and the author of over 30 books on science for the general reader. He has served as an advisor to *Smithsonian Magazine, National Public Radio,* and textbook publishers, as well as many science museums. His writing has been recognized with many awards, most recently the Science Book and Film Editors Award of the AAAS.